Oleg Brovko
Valeri Stepanyuk

Magnetic interactions in nanoscale structures at metallic surfaces

Oleg Brovko
Valeri Stepanyuk

Magnetic interactions in nanoscale structures at metallic surfaces

a first principles study

Südwestdeutscher Verlag für Hochschulschriften

Impressum / Imprint
Bibliografische Information der Deutschen Nationalbibliothek: Die Deutsche Nationalbibliothek verzeichnet diese Publikation in der Deutschen Nationalbibliografie; detaillierte bibliografische Daten sind im Internet über http://dnb.d-nb.de abrufbar.
Alle in diesem Buch genannten Marken und Produktnamen unterliegen warenzeichen-, marken- oder patentrechtlichem Schutz bzw. sind Warenzeichen oder eingetragene Warenzeichen der jeweiligen Inhaber. Die Wiedergabe von Marken, Produktnamen, Gebrauchsnamen, Handelsnamen, Warenbezeichnungen u.s.w. in diesem Werk berechtigt auch ohne besondere Kennzeichnung nicht zu der Annahme, dass solche Namen im Sinne der Warenzeichen- und Markenschutzgesetzgebung als frei zu betrachten wären und daher von jedermann benutzt werden dürften.

Bibliographic information published by the Deutsche Nationalbibliothek: The Deutsche Nationalbibliothek lists this publication in the Deutsche Nationalbibliografie; detailed bibliographic data are available in the Internet at http://dnb.d-nb.de.
Any brand names and product names mentioned in this book are subject to trademark, brand or patent protection and are trademarks or registered trademarks of their respective holders. The use of brand names, product names, common names, trade names, product descriptions etc. even without a particular marking in this works is in no way to be construed to mean that such names may be regarded as unrestricted in respect of trademark and brand protection legislation and could thus be used by anyone.

Coverbild / Cover image: www.ingimage.com

Verlag / Publisher:
Südwestdeutscher Verlag für Hochschulschriften
ist ein Imprint der / is a trademark of
AV Akademikerverlag GmbH & Co. KG
Heinrich-Böcking-Str. 6-8, 66121 Saarbrücken, Deutschland / Germany
Email: info@svh-verlag.de

Herstellung: siehe letzte Seite /
Printed at: see last page
ISBN: 978-3-8381-1627-3

Zugl. / Approved by: Halle, Martin-Luther-Universität Halle-Wittenberg, Diss., 2010

Copyright © 2012 AV Akademikerverlag GmbH & Co. KG
Alle Rechte vorbehalten. / All rights reserved. Saarbrücken 2012

Abstract

Novel possibilities are presented for probing and tailoring magnetic interactions between single adatoms and nanoscale units at metallic surfaces.

Studies, carried out by means of the Korringa-Kohn-Rostoker Green's function method implementation of the density functional theory, give conclusive proofs that it is possible to tune magnetic properties of sub-nanometer structures by adjusting the geometry of the system. The magnetic coupling of a single adatom coupled to a buried magnetic Co layer is shown to attain additional stabilization in either a ferromagnetic or an antiferromagnetic configuration depending on the monolayer's burying depth. It is found that a buried Co layer also has a profound effect on the interatomic exchange interaction between two magnetic impurities on the surface. It is also demonstrated that the exchange interaction between magnetic adatoms at a surface can be additionally manipulated by introducing artificial nonmagnetic Cu chains to link them.

The influence of quantum confinement of surface electrons on the exchange interaction between single adatoms is discussed on the example of a self-assembling system: hexagonal nanoscale paramagnetic islands on a Cu surface. It is demonstrated that it is possible to enhance, reduce or even reverse the exchange coupling at various adatom-adatom separations by deliberate choice of the island's size.

The effect of the exchange interaction on the spin-dependent localization of the surface state is studied by means of *ab initio* calculation. Spin-polarized bound states arising at pairs of magnetic adatoms are shown to be strongly dependent (in terms of both the position and the shape of the bound state peak) on the spin coupling in the system. The possibility is discussed to use the spin splitting of the bound state peak as a novel tool for probing the exchange coupling in the system.

Finally, it is shown that it is not only possible to resolve magnetic properties of small clusters buried beneath a metallic surface, but it is also quite feasible to determine the coupling of buried structures to each other by studying the polarization of electrons in vacuum space above the system.

Abstract in German

Neue Möglichkeiten die magnetische Kopplung in nanometergrossen Oberflächenstrukturen zu untersuchen und zu beeinflussen werden vorgeschlagen und diskutiert.

Anhand der Ergebnisse der Numerischen Simulationen mit der Korringa-Kohn-Rostoker Greensche Funktion Methode [eine Umsetzung der Dichtefunktionaltheorie in der Lokale-Spin-Dichte-Näherung (LSDA)] wird gezeigt, dass man die magnetischen Eigenschaften der Nanostrukturen durch die Anpassung der Systemgeometrie beeinflussen kann. Die Spin-Kopplung einzelner Atome zu, in der Oberfläche versenkten, magnetischen Monolagen wird untersucht und dessen starke Abhängigkeit von der Tiefe der Versenkung bewiesen. Zusätzlich wird gezeigt, dass sich durch bewusste Manipulierung dieser Kopplung die magnetische Wechselwirkung zwischen einzelnen Atomen an der Oberfläche kontrollieren lässt. Eine weitere Möglichkeit der Kontrolle über diese Wechselwirkung, durch Verbinden der einzelnen magnetischen Atome mit paramagnetischen Atomketten, wird angesprochen.

Der Einfluss des Quantenconfinements der Oberflächenzustände auf die Austauschwechselwirkung zwischen einzelnen Atomen wird an dem Beispiel eines selbst assemblierten Systems [hexagonalen Inseln auf der Kupfer Oberfläche] dargestellt. Es wird gezeigt, dass durch eine Anpassung der Inselgrösse die Kopplung magnetischer Atome darauf verstärkt, unterdrükt oder umgekehrt werden kann. Die Untersuchung des Einflusses der interatomaren Austauschwechselwirkung auf die Lokalisierung der Oberflächenzustände ergibt eindeutig, dass die Breite und die energetische Lage der Lokalisierung als ein Indikator der Spin-Kopplung einzelner Atome dienen kann.

Zum Abschluss wird gezeigt, dass sowohl die magnetischen Eigenschaften von den, in die Oberflächen untergetauchten, Nanostrukturen als auch dessen Kopplung zueinander durch die Untersuchung der Polarisierung von Oberflächenelektronnen ermittelt werden können.

Contents

Introduction	1
1 Experimental state of the field	**3**
1.1 Possibilities to control the exchange interaction	3
1.1.1 Surface states as a tool for surface engineering	4
1.1.2 Surface states in quantum corrals	4
1.2 Basics of the STM theory	8
1.3 Existing ways to probe the exchange interaction	10
1.3.1 Probing exchange interaction in linear atomic chains	10
1.3.2 Kondo effect as a spin coupling probe	13
1.3.3 Direct exchange probing by single atom magnetization curves	15
1.4 Theoretical basics	20
1.5 Basic ideas from the theory of chemisorption	21
1.6 Indirect interaction of adatoms: the near zone	24
1.7 Indirect interaction of adatoms: the asymptotic zone	25
1.7.1 Flat Fermi surface	25
1.7.2 Flat sections of the Fermi surface	26
1.7.3 Open sections of the Fermi surface	26
1.7.4 Spherical Fermi surface	26
1.7.5 The role of surface states	26
1.8 Direct interaction	27
1.9 Introducing the magnetism	28
1.9.1 Direct interaction of magnetic impurities	28
1.9.2 Indirect interaction of magnetic impurities	31
Main aims of the study	**32**
2 Theoretical method	**35**
2.1 Density functional theory	35
2.1.1 Hohenberg - Kohn theorems	35
2.1.2 Kohn-Sham (KS) equations	36
2.1.3 The local density approximation	37
2.2 Green's function method	38
2.2.1 Green's functions: definition and basic properties	38
2.2.2 The Dyson equation	39
2.2.3 The Lloyd formula	40
2.2.4 Lippmann-Schwinger equation	41
2.2.5 Single- and multiple-site T-operator	41

		2.2.6 Structural Green's function	42
		2.2.7 Fundamental KKR equation	43
		2.2.8 Spherical representation of the scattering problem	44
		2.2.9 Multiple scattering: bulk crystal	46
		2.2.10 Surfaces and layered systems	48
		2.2.11 Screened KKR	48
		2.2.12 Impurities	49
	2.3	Force theorem	50
	2.4	KKR GF Calculations: workflow and examples	52
		2.4.1 Bulk of an *fcc* copper crystal	52
		2.4.2 Cu(111) surface	55
		2.4.3 A single Co adatom on a Cu(111) surface	56
3	**Tailoring the exchange interaction**		**59**
	3.1	Coupling of single atomic units to a monolayer across a paramagnetic spacer	59
	3.2	Coupling adjustment by linking with nonmagnetic chains	62
	3.3	Confinement on islands as a tool for magnetic coupling control	64
4	**Probing the exchange interaction**		**73**
5	**Magnetism of buried nanostructures**		**81**
	5.1	Probing the electronic and magnetic properties of buried nanostructures	81
		5.1.1 Probing magnetic subsurface impurities with the local density of states at the surface	82
	5.2	Checking the coupling of buried nanostructures	91
	5.3	Outlook	91
References			**95**

Introduction

It has now been long understood, that the "quantum leap" from microelectronics to quantum computing will not be as easy, as the scientific prophets of the middle of the twentieth century have foreboded. Driven to the sub-nanometer limit, the science has faced the fact, that predicting the behavior of nano-scale (purely quantum) systems is profoundly unthinkable without the deepest understanding of all their tiniest properties. Half a century of intensive research has brought solid state science a considerable knowledge harvest, yet there still remain areas which are, to a large extent, insufficiently explored.

This work is dedicated to providing a theoretical coverage for such an unexplored "bare patch" in the fertile field of magnetism and interactions of single sub-nanoscale entities on metallic surfaces, a field which is rather promising for the advancement in the ultimate direction of quantum engineering. The most suitable and experimentally justified basis for nano-magnetism studies is a surface of a crystal. We will try to tackle different aspects of nanostructure magnetism in a system, which is simple enough to allow subsequent generalization of results and at the same time versatile enough to include various fundamental effects: magnetic adatoms and small clusters adsorbed at and below metallic surfaces. The two main things that are necessary to allow to utilize the magnetism of surface structures are the ability to probe the magnetic interaction between single magnetic units and the ability to deliberately adjust this interaction. In the present thesis we particularly focus on the possibility to tailor and probe the exchange interaction of 3d adatoms and clusters at a (111) surface of copper.

To familiarize the reader with the state of the field we will present, in the beginning of the first chapter, the highlights and milestones of the experimental efforts done over the years by the surface science community. We will discuss existing possibilities to manipulate the magnetism of atomic-scale nanostructures and will outline the few feasible approaches to probing of exchange interaction between single atomic units.

One should also not forget, that the theoretical foundations for the description of the magnetism and interaction of single atoms (impurities) have been laid more than 60 years ago. In the second part of the first chapter we will shortly discuss the theoretical basics of Anderson and Friedel theories as well as modern advancements in the description of impurity interaction at surfaces. Having a sound theoretical foundation is essential for justification and assessment of the results of the calculation done in the framework of the present thesis.

Details of the calculation technique involved are presented in Chapter 2. The chapter starts with an introduction to the basics of the density functional theory where the substitution of a many-particle quantum problem by an equivalent single-particle one is briefly discussed. The main part of the chapter is devoted to a thorough, yet focused, description of the Korringa-Kohn-Rostoker's Green's function method, which is the main tool, used in the present study. Several essential theorems are introduced. The chapter is concluded by a sample of a calculational workflow, exemplified by a calculation of electronic and magnetic properties of a single Co adatom adsorbed on a Cu(111) surface.

In Chapter 3 we discuss several novel possibilities to control the exchange interaction in sub-nanoscale structures. We give conclusive proofs that it is possible to tune magnetic properties of sub-nanometer structures by adjusting the geometry of the system. The magnetic coupling of a single adatom to a buried magnetic Co layer is shown to attain additional stabilization in either a ferromagnetic or an antiferromagnetic configuration depending on the monolayer's burying depth. It is found that a buried Co layer also has a profound effect on the interatomic exchange interaction between two magnetic impurities on the surface. Additionally it is demonstrated that the exchange interaction between magnetic adatoms on a surface can be manipulated by introducing artificial nonmagnetic chains to link them.

The influence of quantum confinement of surface electrons on the exchange interaction between single adatoms is discussed in the final part of Chapter 3 on the example of a self-assembling system: hexagonal nanoscale paramagnetic islands on a Cu surface. It is demonstrated that it is possible to enhance, reduce or even reverse the exchange coupling at various adatom-adatom separations by deliberate choice of the island's size.

Chapter 4 is aimed at convincing the reader, that the exchange interaction has a profound influence on the spin-dependent localization of the surface state. Spin-polarized bound states arising at pairs of magnetic adatoms are shown to be strongly dependent (in terms of both the position and the shape of the bound state peak) on the spin coupling in the system. The possibility is discussed to use the spin splitting of the bound state peak as a novel tool for probing the exchange coupling in the system.

Finally, in Chapter 5, it is shown that it is not only possible to resolve magnetic properties of small clusters buried beneath a metallic surface, but it is also quite feasible to determine the coupling of buried structures to each other by studying the polarization of electrons in vacuum space above the system.

The thesis is then underlined by a short summary and a conclusion.

Chapter 1

Magnetic interaction at the atomic-scale: experimental state of the field and theoretical basics

The foundations of spin engineering were laid by P. Grünberg and A. Fert [1, 2] with the discovery of the giant magnetoresistance (GMR) effect. Of course, they did neither discover the spin nor any new ground-braking properties of it, but they have clearly shown the world, that it is possible to employ the microscopic magnetic properties of matter in our everyday engineering tasks. Since that discovery, the direction of many new solid state studies has been predefined by the challenges in the field of spintronics [3]. One of the main challenges is the need to gain meticulous control of the magnetism of single atomic-scale units. Such control is, however, unthinkable without a deeper understanding of the manner in which this units interact with each other. This understanding can, in turn, only be obtained if we develop an adequate set of tools to probe and manipulate this interaction. In the present chapter we will try to outline the already existing possibilities to achieve that.

1.1 Possibilities to control the exchange interaction

Above the nanoscale, the possibility to couple single spin has been extensively studied in various, lithographically fabricated, quantum dots [4, 5]. At the atomic length scale single spins can be coupled directly in metal-atom clusters [6, 7] and chains [8]. Yet such techniques largely lack the flexibility and versatility.

Quite recently it has been shown, that it is possible to directly affect single spins on the surface by means of a scanning tunneling microscope (STM) tip (Fig. 1.1) [9]. The exchange coupling between single 3d magnetic adatoms (Cr, Mn, Fe, and Co) adsorbed on a Cu(001) surface and a Cr STM tip has been shown to be controllable by the variation of the tip-substrate distance [9]. The sign of the exchange energy was found to be determined by the competition of the direct and the indirect interactions between the tip and the adatom [9]. This exchange control method has, however, the disadvantage, that such a vacuum tunneling junction requires a complex apparatus of an STM to be created. In Section 3.1 we will show, that coupling of single spins to extended structures (like an STM tip or a magnetic layer) can be much more effective through a paramagnetic spacer, similar to the case of the famous interlayer exchange coupling (IEC) [1, 10, 11, 12]. Such coupling might allow one to conveniently stabilize or tailor single spins or spin groups at the surface.

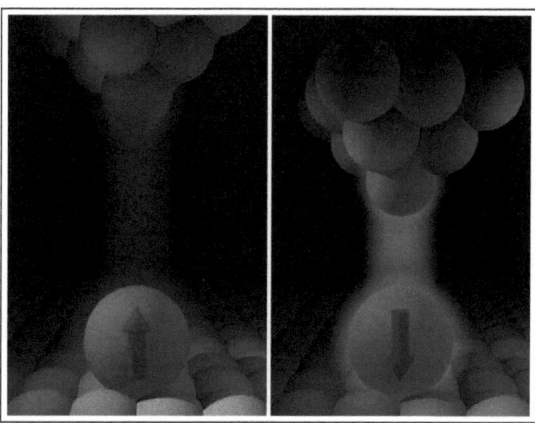

Figure 1.1: The spin orientation of single 3d magnetic adatoms adsorbed on a Cu(001) surface can be controlled by an STM with a magnetic (Cr) tip by varying the tip-substrate distance [9].

1.1.1 Surface states as a tool for surface engineering

Another tool for surface studies and engineering, which can be used to control the exchange coupling of single atoms at surfaces and which has few rivals in elegance, versatility and efficiency, are the surface state electrons. Surface state is a two-dimensional confinement of electrons between the vacuum barrier and the inverted band-gap of the bulk which exists at some surfaces of some noble metals (see Subsection 1.7.5 for details). The scattering of surface state electrons at point defects [13, 14], single adsorbates [15, 16] and extended structures [17, 18, 19, 20] leads, due to high correlation and the intrinsic two-dimensionality of the surface state, to the formation of interference patterns which manifest themselves as standing waves in the local density of electronic states (LDOS). These standing waves can be detected by various surface sensitive techniques such as the scanning tunneling (STM) microscopy (see the next section for details). Standing LDOS waves can induce palpable changes in adsorption and activation energies of surface impurities and are the cause of an indirect long-range interaction (LRI) between them as is discussed in detail in Section 1.7.5 and in [21, 22, 23, 24, 25, 26, 27, 28].

1.1.2 Surface states in quantum corrals

At some point in the history of the surface science development it was realized that by deliberate arrangement of atomic scatterers [29, 30] one can confine surface state electrons to close geometries thus precisely tailoring their interference.

In 1994 Heller et al. [29] presented a combined theoretical and experimental description of the surface-state confinement to a *quantum stadium* built of 76 Fe atoms on a Cu(111) surface [29]. Heller and coworkers used a simple theoretical model, an advanced version of a particle-in-a-box model, to describe the confinement of electrons theoretically. Experimental

1.1 Possibilities to control the exchange interaction

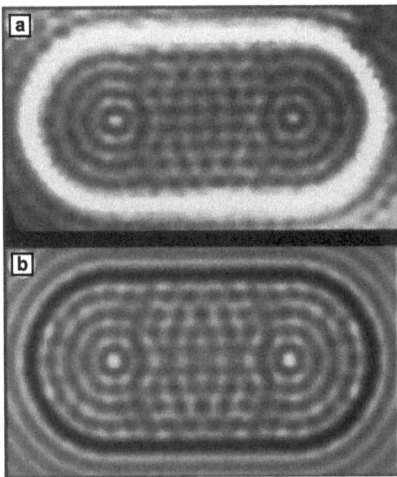

Figure 1.2: LDOS near E_F for a 76 atom "stadium" of dimensions 141×285 Å: experiment (a), bias voltage 0.01 V, and theory (b). The LDOS near the center of Fe adatoms is not accounted for in the theory and appears black. Figure from [29].

and theoretical local density of states (LDOS) at the Fermi level in this corral are presented in Fig. 1.2.

A decade later Manoharan and coworkers demonstrated that quantum corrals could be exploited to project electronic structure of adatom onto remote location [30]. They assembled an elliptical Co corral on Cu(111) (Fig. 1.3a,b) surface and placed into one focus of the ellipse a magnetic Co atom (Fig. 1.3c). It is a matter of simplest geometry that circular a wave emitted from one focus of an ellipse is focused in the other one. What was surprising was that the surface state electrons of Cu(111) confined to elliptical quantum corral obey the same rules [30]. Manoharan and coworkers could observe the electronic structure of a Co adatom being coherently refocused onto the empty focus spot [30]. A distinctive spectroscopical feature, known as many-particle Kondo resonance, was used to probe this effect [31, 32]. The Kondo resonance, a signature of magnetic moment coupled to the sea of conduction electrons, was found in an empty focus (see Fig. 1.3d). This finding was patented as a method for information transfer and replication between spatially distinct points via the engineering of quantum states [US Patent 6441358].

Experiments of Manoharan et al. [30] on quantum mirages in quantum corrals inspired a number of theoretical works. Ab initio calculations performed by means of the KKR method for the elliptical corrals used by Manoharan and coworkers in their experiments unambiguously demonstrated that the spin polarization of the surface-state electrons caused by magnetic adatoms can be projected to a remote location and can be strongly enhanced in corrals, compared to an open surface (Fig. 1.3) [33].

Adjustable geometries of quantum corrals open up wide range of possibilities, such as the ability to affect the motion of single atoms [34, 35] or the ability to engineer electronic states [36, 37]. But one possibility will be of particular interest to us - the possibility to

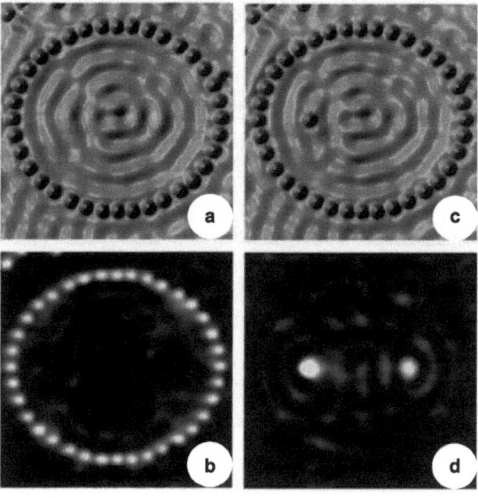

Figure 1.3: (a) Topographs of an elliptical corral with eccentricity $e = 1/2$ (the larger axis $a = 71.3$ Å) employing Co atoms to corral two-dimensional electrons on Cu(111). (b) Associated spectroscopy maps. (c) Topographs showing the corral with a Co atom at the left focus and (d) associated spectroscopy maps showing the Kondo effect projected to the empty right focus [30], resulting in a Co atom mirage. Image dimensions are 150 × 150 Å. [30]

1.1 Possibilities to control the exchange interaction

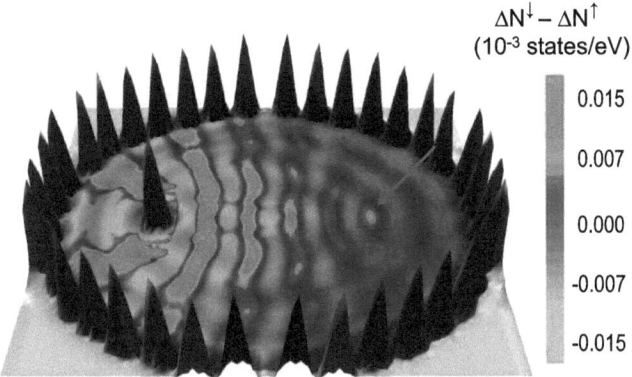

Figure 1.4: The LDOS at the Fermi energy on the Co adatom and the Co atoms of the corral walls are shown. The spin polarization of surface-state electrons inside the Co corral is presented in color: $\Delta N^{\downarrow}$ and $\Delta N^{\uparrow}$ are determined by the difference between LDOS near the Fermi energy (+10 meV) of the Co corral with the Co adatom, the empty Co corral, and the single Co adatom on the open Cu(111). The mirage in the empty focus is marked by the red arrow. The geometrical parameters of the corral are the same as in the experimental setup of Ref. [30], i.e., ellipse semiaxis $a = 71.3$ Å and eccentricity $e = 0.5$. The present figure is taken from [33]

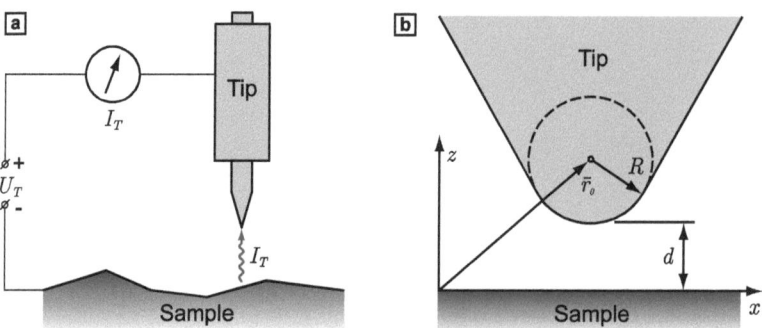

Figure 1.5: (a) The principle of STM. A tunneling voltage is applied between the sample and a scanning probe. The tunneling current is registered. (b) Tip-sample geometry approximation of Tersoff and Hamann [48, 49].

tailor the interatomic exchange interaction between magnetic adatoms at large separations by adjusting the corral geometry has been proven by Niebergall et al. [34]. A wide application of this possibility is hindered, to some extent, by the tremendous effort needed to construct such complex nanostructures in an atom-by-atom fashion. In Section 3.3 of the present thesis we demonstrate, that one can utilize the same effect of quantum confinement in self assembling structures to gain the same amount of control over the the coupling between single atomic spins.

Even having found an ultimate way of deliberately tailoring the magnetic interaction between single atomic units, we will still not be able to make any fundamental progress without having an adequate technique for probing the interaction we have just tailored.

Spin interactions in surface systems have typically been probed using ensemble-averaging techniques, such as susceptibility measurements [38, 39], electron paramagnetic resonance [40] or inelastic neutron scattering [41]. All of them being fundamentally non-local (on the nanoscale) techniques, even achieving a truly atomic resolution in studying nanostructures at surfaces has for a long time been a dream of solid-state physics scientists. This dream came true with the invention of the scanning tunneling microscopy (STM) technique by Binnig and Rohrer [42, 43]. The technique utilized the extreme distance dependence of the electron tunneling to probe geometric properties of surface structures with atomic resolution [44]. The combination of the intrinsic locality of the technique with electron tunneling spectroscopy methods has lead to the appearance of the scanning tunneling spectroscopy (STS) - a powerful method for studying the electronic structure of the surface resolved at the nanoscale. Soon after the original discoveries it was realized, that by using magnetic probe tips one could also access the information about the spin character of the electronic states of the system - the, so called, spin-polarized STM and STS techniques [45, 46, 47].

1.2 Basics of the STM theory

The key value, measured in STM and STS, is the tunneling current or, alternatively, the conductance between the sample nanostructure and the tip of the probe (Fig. 1.5a). The

1.2 Basics of the STM theory

link between this value and the actual electronic properties of the sample (particularly its electronic states) has been made by Tersoff and Hamann [48, 49]. The Tersoff-Hamann model is a specialization of the perturbative treatment of planar tunneling, introduced originally by Bardeen [50]. The tunneling current has been derived by Bardeen in the formalism of the first-order time-dependent perturbation theory:

$$I = \frac{2\pi e}{\hbar} \sum_{\mu,\nu} \{f(E_\mu)[1 - f(E_\nu + eU_T)] - f(E_\nu + eU_T)[1 - f(E_\mu)]\} \cdot |M_{\mu\nu}|^2 \delta(E_\nu - E_\mu), \quad (1.2.1)$$

where U_T is the applied bias voltage, $f(E)$ is the Fermi function, $M_{\mu\nu}$ is the element of the tunneling matrix between the unperturbed electronic states Ψ_μ of the tip and Ψ_ν of the surface. $E_{\mu(\nu)}$ are the energies of corresponding states in absence of the tunneling. The elastic tunneling and the corresponding energy conservation are described by the δ function. In a more general representation the matrix element $M_{\mu\nu}$ in Bardeen's formalism is given by

$$M_{\mu\nu} = -\frac{\hbar^2}{2m} \iint \mathbf{dS} \left(\Psi_\mu^* \nabla \Psi_\nu - \Psi_\nu^* \nabla \Psi_\mu \right), \quad (1.2.2)$$

where the integral has to be evaluated over any surface lying entirely within the vacuum barrier separating the two electrodes. The quality being integrated can be viewed as a density of the current $\overline{j}_{\mu\nu}$. So the main problem of solving (1.2.2) is finding adequate explicit representations for the tip and the surface wave functions. Tersoff and Hamann [48, 49] proposed a simple approximation of a spherically symmetric tip (Fig. 1.5). A corresponding wave function for the evaluation of the matrix element can be written for a tip with a curvature radius R as an s-type wave function with en effective decay length κ:

$$\Psi_\mu = \frac{1}{R} e^{-\kappa R}. \quad (1.2.3)$$

Furthermore, Tersoff and Hamann have assumed the limit of low temperatures and small bias voltages to get rid of Fermi distributions in (1.2.1) and rewrite the tunneling current as follows:

$$I_T = \frac{2\pi e}{\hbar} U_T \sum_{\mu,\nu} |M_{\mu\nu}|^2 \delta(E_\mu - E_F)\delta(E_\nu - E_F). \quad (1.2.4)$$

Then, within the s-wave tip approximation (1.2.3), the tunneling current can be further rewritten as:

$$I_T \propto U_T n_t(E_F) e^{2\kappa R} \sum_\nu |\Psi_\nu(\overline{r}_0)|^2 \delta(E_\nu - E_F), \quad (1.2.5)$$

where $n_t(E_F)$ is the density of states at the Fermi level and $\overline{r}_0$ is the center of curvature of the tip. If we look at the last to terms of (1.2.5):

$$n_s(E_F, \overline{r}_0) = \sum_\nu |\Psi_\nu(\overline{r}_0)|^2 \delta(E_\nu - E_F), \quad (1.2.6)$$

we can identify them as the local density (LDOS) of sample states evaluated at the position of the tip. Thus it can be concluded that *the tunneling current in a STM is roughly proportional to the LDOS of the sample*, which allows one, among other things, to compare experimental results with those, obtained in theoretical calculations.

If either the tip of the sample (or both) are made of a magnetic material, the tunneling current acquires an additional spin dependance. In the limit of the vanishing bias voltage it can be shown [51, 52] that the tunneling current can be expressed as:

$$I_T(\bar{r}_0, U, \theta) = I_0(\bar{r}_0, U_T) + I_{sp}(\bar{r}_0, U_T, \theta) =$$
$$= \frac{4\pi^3 C^2 \hbar^3 e}{\kappa^2 m^2} \left[n_t \tilde{n}_s(\bar{r}_0, U_T) + \overline{m}_t \overline{m}_s(\bar{r}_0, U_T) \right], \quad (1.2.7)$$

where n_t is the non-spin-polarized LDOS of the tip, $\tilde{n}_s$ is the energy-integrated LDOS of the sample, and $\overline{m}_t$ and $\overline{m}_s$ are the corresponding magnetization vectors of the spin-polarized LDOS:

$$\overline{m}_s(\bar{r}_0, U_T) = \int^{eU_T} \overline{m}_s(\bar{r}_0, E) \mathrm{d}E \quad (1.2.8)$$

where

$$\overline{m}_s = \sum_\mu \delta(E_\mu - E)\, \Psi_\mu^{S\dagger}(\bar{r}_0)\, \sigma\, \Psi_\mu^S(\bar{r}_0). \quad (1.2.9)$$

Ψ_μ^S is the spinor of the wave function

$$\Psi_\mu^S = \begin{pmatrix} \Psi_{\mu\uparrow}^S \\ \Psi_{\mu\downarrow}^S \end{pmatrix} \quad (1.2.10)$$

and σ is the Pauli's spin matrix. It is important to note, that the spin-polarized part of the current is being scaled with the projection of $\overline{m}_s$ onto $\overline{m}_t$ [52]. Experimentally, this technique was implemented for the first time by Wiesendanger *et al.* [45]. Later, an alternative mode of SP-STM operation, aiming at a separation of electronic and magnetic structure information, was introduced at a later time by Wulfhekel and Kirschner [53]. SP-STM technique, providing an unprecedented insight into both magnetic and non-magnetic atomic-scale properties of surfaces alike, has long since become one of the main techniques for surface studies and the possibility of a direct link to theoretical results has provided an invaluable support for many theoretical studies, including the present one.

1.3 Existing ways to probe the exchange interaction

Having such an extraordinary tool as the STM, even with a truly atomic resolution, does not, unfortunately, provide us with a recipe of accessing the information about the coupling between single atomic spins. And until recently, such direct measurement of the magnetic interaction has been, indeed, impossible. First steps in this field have been made by Hirjibehedin and co-workers as they managed to experimentally probe the spin exchange interaction in linear manganese chains through spin-flip experiments [54].

1.3.1 Probing exchange interaction in linear atomic chains

Hirjibehedin and coworkers [54] used a scanning tunneling microscope to probe the interactions between spins in individual atomic-scale magnetic structures. They have combined the imaging and manipulation [55] capabilities of the STM with its ability to measure spin excitation spectra at the atomic scale [56]. They used an STM to build magnetic chains up to 10 Mn atoms long on a thin insulating copper nitride (CuN) island (Fig. 1.6A) and probed

1.3 Existing ways to probe the exchange interaction

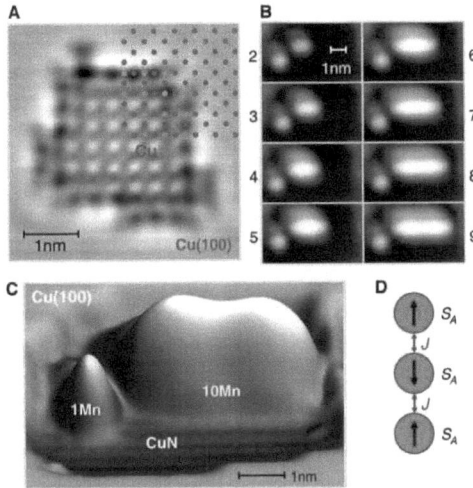

Figure 1.6: Mn chains on CuN. (A) STM constant-current topograph (10 mV, 1 nA) of a CuN island on Cu(100). Topograph is high-pass (curvature) filtered to enhance contrast, with lattice positions of Cu (red dots) and N (blue dots) atoms overlaid. CuN islands appear as 0.14 nm depressions for $|V| < 0.1$ V. (B) STM images of the building of a chain of Mn atoms, lengths 2 to 9, on CuN (10 mV, 0.1 nA). Individual atoms in the chain cannot be resolved. Artifacts seen to the upper left of all structures are characteristic of the atomic arrangement of the tip used for manipulation. (C) Perspective rendering of a chain of 10 Mn atoms. (D) Schematic of the antiferromagnetic coupling of three atomic spins described by the Heisenberg model in Eq. (1.3.1). Figure taken from [54].

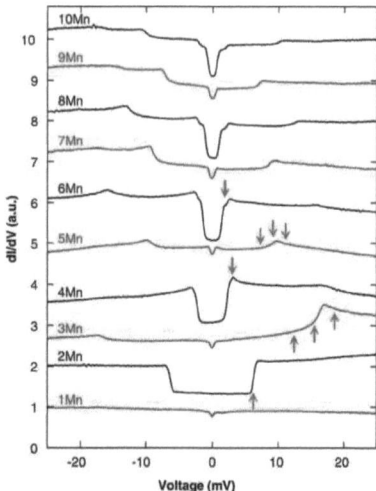

Figure 1.7: Conductance spectra of Mn chains on CuN. Spectra were taken with the tip positioned above the center of chains of Mn atoms of lengths 1 to 10 at $T = 0.6$ K and $B = 0$ T. Spectra were acquired at a nominal junction impedance of 20 megohms (20 mV, 1 nA) and were not sensitive to junction impedance. Successive spectra are vertically offset by one unit for clarity. Odd spectra are in red and even spectra are in black to emphasize the parity dependence. Blue arrows indicate the lowest energy spin-changing excitation obtained from the Heisenberg model described in Eq. (1.3.1) with $J = 6.2$ meV and $S_A = \frac{5}{2}$. For comparison, green arrows indicate the same excitation with $S_A = 2$ (lower voltage) and $S_A = 3$ (higher voltage) on the trimer and pentamer; the same excitation with $S_A = 2$ and $S_A = 3$ is not distinguishable from $S_A = \frac{5}{2}$ on the dimer, tetramer, and hexamer chains. From [54].

their collective spin excitations with inelastic electron tunneling spectroscopy (IETS). The insulating properties of a single atomic layer of CuN were needed to screen atomic spins from the Cu(100) surface [54].

The resulting dI/dV spectra, taken above the middles of Mn chains of 1 to 10 atoms at $B = 0$ (Fig. 1.7) displayed a striking dependence on the parity of the chain. At voltages below 1 mV, the spectra of odd-length chains, including the single Mn atom, exhibited a narrow dip in conductance that was centered at 0 V. This feature was absent in all even-length chains. At larger energies, both odd and even chains, with the exception of single Mn atoms, showed large steps in conductance at voltages that were symmetric with respect to zero [54].

Hirjibehedin and colleagues [54] have used IETS notions to interpret the steps in the dI/dV spectra of the chains. When a system with a mode at energy E is placed within a tunnel junction, electrons can excite this mode during the tunneling process only if the magnitude of the bias voltage V exceeds E/e (where $-e$ is the charge on the electron). This additional tunneling channel generally results in steps in conductance at $V = \pm E/e$ [54].

The absence of the spin-flip excitation on the even-length chains of an antiferromagnetic (AF) order (as is Mn) is consistent with the fact the an antiferromagnetic even-length chain has a ground-state spin of $S = 0$. The odd-length chains, on the contrary have a net spin of $S = 1$ in the ground state and thus exhibit low energy spin-flip excitations [54].

For the even-length chains in Fig. 1.7 the lowest energy conductance steps were interpreted as excitations that change the total spin of the chain [54]: a spin-changing transitions from the ground state (with $S = 0$ and $m = 0$) to an excited state (with $S = 1$ and $m = -1, 0, 1$ (degenerate at $B = 0$)).

The analytical interpretation of spectroscopic results was done by using the Heisenberg model for an open chain of N coupled spins in its simplest form. It included only identical nearest neighbor exchange interactions of strength J between the spins (Fig. 1.6D). All sites were assumed to have the same spin S_A. The Hamiltonian had the form of [54]:

$$H_N = J \sum_{i=1}^{N-1} S_i \cdot S_{i+1}, \qquad (1.3.1)$$

where S_i is the spin operator for the i-th site along the chain. The eigenvalue of S_i^2 was thus $S_A(S_A + 1)$ for every i.

For a dimer, the ground state of the AF ($J > 0$) Heisenberg chain is a singlet ($S = 0$) and the first excited state is a triplet ($S = 1$) with energy separation J [54]. This result is independent of the spin S_A of the constituent atoms. The energy spacing between the ground state and the first excited state provides a direct measure of the coupling strength J [54]. For the dimer shown in Fig. 1.7, the determined coupling was $J = 6.2$ meV.

In contrast to the dimer, the ground state of the Heisenberg AF trimer has total spin $S = S_A$. Generally, the first excited state has total spin $S_A - 1$ and is higher in energy by $S_A \cdot J$ [54]. Figure 1.7 shows the energy of the first excited state for the trimer for a few possible values of S_A, where the value of J used was that determined from the dimer. The conductance step at ~ 16 mV, which can be now assigned to the lowest spin-changing transition, is best matched with the result for $S_A = \frac{5}{2}$.

From the excitation spectra of the dimer and trimer, Hirjibehedin and coworkers [54] determine both free parameters J and S_A of the Heisenberg Hamiltonian in Eq. (1.3.1). Using these two values, they then calculate the excitation spectra for chains up to $N = 6$. For odd-length chains, they find that the ground and first excited states have total spin $\frac{5}{2}$ and $\frac{3}{2}$, respectively. In contrast, even-length chains have ground and first excited state spins of 0 and 1, respectively. This result is consistent with the observed parity dependence of the spin-flip feature and with the triplet splitting of the spin-changing excitation of even-length chains. The fact the the calculated energy of the first excited state, as shown in Fig. 1.7 is in very good agreement with the observed spectral features proves once again, that the described procedure allows one to deduce the nearest neighbor exchange coupling parameters from the IETS spectra.

1.3.2 Kondo effect as a spin coupling probe

Another method, based on the analysis of the Kondo resonance, was proposed by Chen *et al.* for Cobalt dimers on a Au(111) surface [57] and successfully extended and improved by Wahl *et al.* for Co on Cu(111) [58].

The Kondo effect is a collective electronic resonance which originates from the screening of the spin of a magnetic impurity by the surrounding conduction band electrons [59]. It

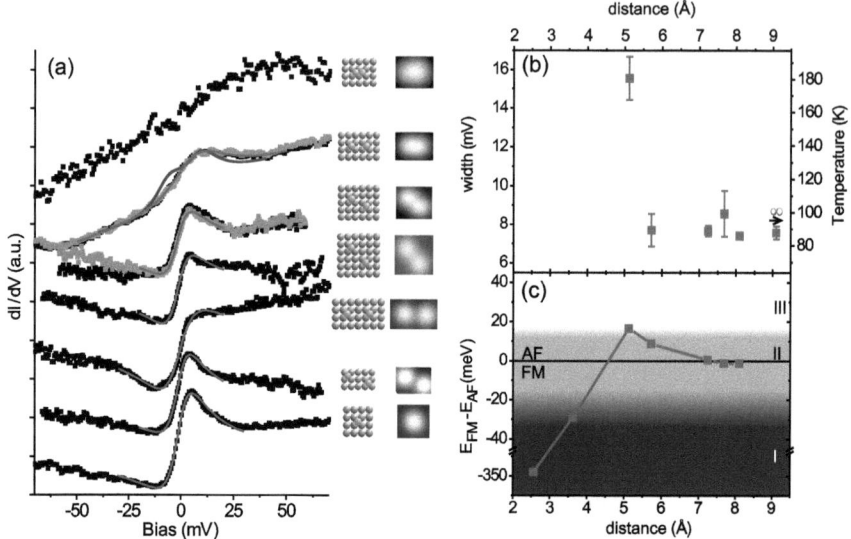

Figure 1.8: Kondo resonance of cobalt dimers on Cu(100) measured by STS at 6 K. As a consistency check, spectra taken on both ends of the dimers are shown (green and black dots) to be equivalent. (a) Model, topography, and spectra for (from top to bottom) a compact dimer (2.56 Å), a dimer at 5.12, at 5.72, at 7.24, at 7.68, at 8.10 Å, and for a single adatom at infinite distance (> 20 Å) are depicted. The spectra are shown together with fits by a Fano function (red solid line), for the dimer at 5.12 Å also a simulated curve with $J = 15$ meV and $\Gamma = 1.2\,T_K^0$ is plotted (blue solid line). (b) The width of the resonance as a function of distance, and (c) KKR calculations for the exchange interaction between cobalt adatoms on Cu(100) [63]. The three distinct regimes discussed in the text are shaded in different grays. Figure from [58].

is characterized by a strong peak in the impurity's density of states near the Fermi level. Kondo effect has been observed for various surface systems: single magnetic adatoms [31, 32], artificial nanostructures [30], and for molecules [60]. In STS spectra, it shows up as a feature which can be described by a Fano line shape [61].

From a fit, the peak width Γ is obtained which is the characteristic energy scale (the Kondo temperature T_K) of the impurity system [58]. For the Kondo scenario of a single magnetic impurity on a nonmagnetic metal surface a quantitative description has been proposed [62]. A proximity of a second impurity might change the picture drastically.

Chen and coworkers [57] were the first to notice, that the exchange coupling in magnetic dimers has a notable influence on the Kondo screening of the system. They did, however, only observe the disappearance of the Kondo resonance in nearest neighbor dimers on Au(111) and did not give any recipe for extracting the amplitude of the magnetic interaction between single impurities.

In their paper [58], Wahl and coworkers demonstrate that it is possible to determine the magnetic interaction between single magnetic atoms adsorbed on a noble metal surface by measuring the modified Kondo spectrum. They measure the evolution of the Kondo line shape obtained by STS upon varying the interatomic distance between Co adatoms in dimers on a Cu(100) single crystal surface. The panels in Fig. 1.8a show different cobalt dimer configurations prepared in the experiment [58] with interatomic distances between the neighboring cobalt atoms ranging from 2.56 to 8.1 Å together with the corresponding STS spectra. For the compact dimer the interaction between the spins is much stronger than the coupling to the substrate and the Kondo effect (at 6 K) is suppressed [58]. For the next-nearest neighbor distance, however, a resonance is found at the Fermi energy. The resonance is considerably broader than that of a single cobalt adatom. By fitting the STS signal with a single Fano line shape, Wahl et al. state that the energy width of the feature would correspond to a Kondo temperature $T_K = 181 \pm 13$ K [58]. For distances of 5.72 and 7.24 Å, the Kondo resonance has already recovered almost the same width and line shape as that of a single cobalt adatom. For even larger distances, the same width as on an adatom is restored. The widths of the resonances are summarized in Fig. 1.8b [58].

The authors interpreted their data theoretically as a realization of a two-impurity Kondo problem [64]. Depending on the relative strength of the exchange interaction compared to the single-impurity Kondo temperature T_K^0, the dimers enter different regimes. For a strong ferromagnetic exchange interaction $|J| \gg T_K^0$ (marked as regime I in Fig. 1.8c) a correlated state with a new Kondo temperature $T_K^{dimer} \approx (T_K^0)^2/|J|$ will occur [64]. This new Kondo scale is much lower than the temperature of the experiment and can therefore not be detected in direct measurements [58]. For intermediate exchange interaction J (regime II in Fig. 1.8c), the single-impurity Kondo resonance is recovered. Finally, for a sufficiently strong antiferromagnetic exchange interaction $J > J^* \sim 2T_K^0$ (marked as regime III in Fig. 1.8c) between neighboring magnetic atoms, the Kondo resonance is split and a singlet state is formed between the impurities [64]. This singlet state is characterized in the impurity density of states by peaks located at energies $\pm J/2$ [65, 66, 58].

Using the slave boson mean field theory approach from Ref. [66] Wahl and coworkers approximate the split Fano curve in the resulting density of states by the following formula:

$$\rho(\epsilon) \propto a_1 f\left(\frac{\epsilon + J/2}{\Gamma}, q\right) + a_2 f\left(\frac{\epsilon - J/2}{\Gamma}, q\right), \qquad (1.3.2)$$

where $f(x,q) = \frac{(x+q)^2}{x^2+1}$ and $a_1 \sim a_2$. This equation describes two Fano resonances at $\pm J/2$ with the same width Γ which is of the same order as the single-impurity one [58]. The resonances are resolved in the tunneling spectrum as only one broadened feature as observed in the experiment due to the width of the resonances, which is of the same order as the splitting.

Thus, the authors show, that the width of the resonance is a direct measure for the magnetic interaction between the adatoms [58]. The exchange interaction values obtained by fitting the spectra in Fig. 1.8a with function (1.3.2) have been corroborated by Korringa-Kohn-Rostoker Green's function calculations. KKR calculated values for the exchange interaction between cobalt adatoms on Cu(100) are shown in Fig. 1.8c [63, 58].

1.3.3 Direct exchange probing by single atom magnetization curves

And quite recently a state of the art experiment by Meier et al. allowed to directly map the exchange coupling of single adatoms with the atoms' magnetic response to an external

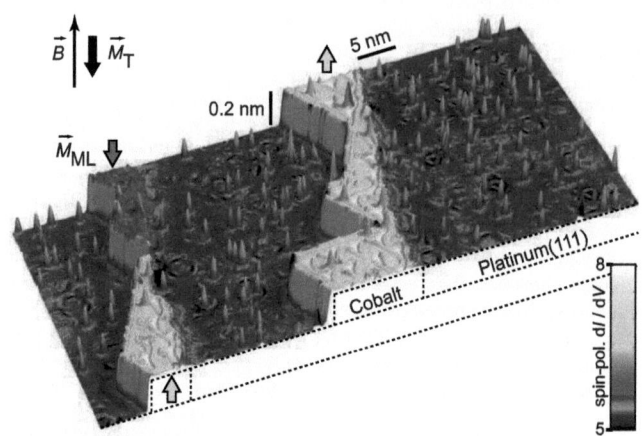

Figure 1.9: Overview of the sample of individual Co adatoms on the Pt(111) surface (blue) and Co ML stripes (red and yellow) attached to the step edges (STM topograph colorized with the simultaneously recorded spin-polarized dI/dV map measured with an STM tip magnetized antiparallel to the surface normal). An external magnetic field $\overline{B}$ can be applied perpendicular to the magnetization of adatoms $\overline{M}_A$, ML stripes $\overline{M}_{ML}$ or the tip $\overline{M}_T$. The ML appears red when $\overline{M}_{ML}$ is parallel to $\overline{M}_T$ and yellow when $\overline{M}_{ML}$ is antiparallel to $\overline{M}_T$. (Tunneling parameters are as follows: $I = 0.8$ nA, $V = 0.3$ V, modulation voltage $V_{mod} = 20$ mV, $T = 0.3$ K). From [67].

magnetic field (magnetization curves) [67].

In their experiment, Meier and coworkers study the magnetization of single Co adatoms deposited at about 25 K on the bare stepped Pt(111) and their interaction with stripes of one atomic layer Co grown at room temperature at the step edges (Fig. 1.9) [67].

Using the SP-STM technique they record spin-polarized differential conductance spectra above single Co adatoms and present them decomposed as [67]

$$dI/dV \propto (dI/dV)_0 + (dI/dV)_{SP}\left(\overline{M}_T \cdot \overline{M}_A\right), \qquad (1.3.3)$$

where the first term represents the bias-voltage-dependent spin-averaged part of the dI/dV and the second term introduces the spin-resolved differential conductance. dI/dV (averaged over a time window) as a function of an external magnetic field, $\overline{B}$, is thus directly proportional to the time-averaged magnetization $\langle \overline{M}_A \rangle$ in the out-of-plane direction [67].

Considering that the magnetization of the tip stays constant (confirmed experimentally [67]) the authors then apply an external magnetic field and, varying it's strength, obtain the dependence of the dI/dV signal on $\overline{B}$. Such a dependence is presented in Fig. 1.10 for two different temperatures of the system. The authors then extract the value of the atom's magnetic moment by fitting the curves by using the magnetic energy function [68, 67]

$$E(\theta, B) = -m(B - B_T)\cos\theta - K(\cos\theta)^2,$$

1.3 Existing ways to probe the exchange interaction

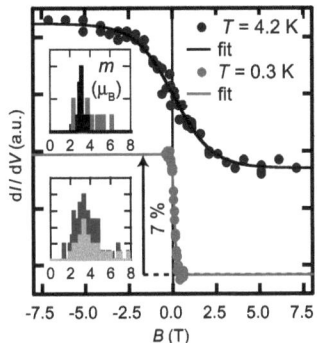

Figure 1.10: Magnetization curves from the same adatom taken at different temperatures as indicated (dots). The solid lines are fits to the data (see text). The insets show the resulting histograms of the fitted magnetic moments (in μ_B) for the same 11 adatoms at $T = 4.2$ K (black) and at 0.3 K (red) (top histogram) and for 38 hcp (orange) and 46 fcc (blue) adatoms at 0.3 K (bottom histogram, fcc bars stacked on hcp). Tunneling parameters are as follows: $I = 0.8$ nA, $V = 0.3$ V, $V_{mod} = 20$ mV. From [67].

where θ is the angle between the magnetic moment and $\overline{B}$. The authors vary the magnetic moment m, the saturation magnetization M_{sat}, the tip stray field B_T and use the anisotropy value K as given in [68].

Similar magnetization curves as in Fig. 1.10 have been recorded by using several tips for about 80 different adatoms showing qualitatively the same paramagnetic shape. The insets contain the histograms of the fitted m. The increased variance at $T = 0.3$ K can, according to Meier et al. [67], only be accounted for if one considers the atoms to be interacting with the rest of the system. Moreover, the only interaction, which could account for such a large dispersion of m was the substrate mediated long range interaction between the adatoms and the magnetic stripes. The assumption was corroborated by the fact, that the magnetic behavior of atoms was strikingly different in the vicinity of the stripes [67].

Focusing on adatoms close to the ascending edge of the stripe, the authors [67] measure the magnetization curve of a particular ML stripe and of an adatom with a distance of $d \approx 1.5$ nm (Fig. 1.11). They describe the behavior of the system as follows [67]. The ML shows a regular square-like hysteresis corresponding to ferromagnetic behavior [67]. In the down sweep (blue curve), its magnetization switches from up (high signal) to down (low signal) at $B = -0.5$ T, and in the up sweep (red curve) it switches from down to up at $+0.5$ T. The adatom close to a stripe behaves completely different than the distant ones (which showed paramagnetic behavior). In the down sweep, its magnetization switches from up to down already at large positive $B = +0.7$ T (see arrow). It then switches back to up simultaneously with the reversal of the stripe at -0.5 T. Only at -0.7 T is the adatom's magnetization again forced into the down state (see arrow). The same behavior is observed for the up sweep but now with the stripe magnetization pointing downward and the adatom magnetization pointing upward at zero field. The adatom feels an anti-ferromagnetic (AF) coupling to the stripe, which is broken by an exchange bias of $B_{ex} = \pm 0.7$ T corresponding to an interaction energy of $J = -m \cdot B_{ex} \approx -150$ µeV ($m = 3.7$ μ_B). The magnetization

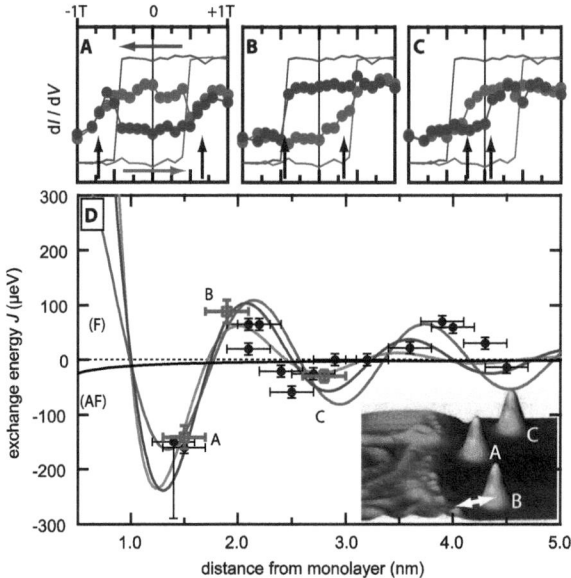

Figure 1.11: Magnetic exchange between adatoms and ML stripe. (A to C) Magnetization curves measured on the ML (straight lines) and on the three adatoms (dots) A, B, and C visible in the inset topograph of (D). The blue color indicates the down sweep from $B = +1$ T to -1 T (and red, the up sweep from -1 T to $+1$ T) (dI/dV signal on ML inverted for clarity). The vertical arrows indicate the exchange bias field, B_{ex}, which is converted into the exchange energy (using $m = 3.7\mu_B$) for the corresponding magenta points in the plot (D). (Tunneling parameters are as follows: $I = 0.8$ nA, $V = 0.3$ V, $V_{mod} = 20$ mV, $T = 0.3$ K.) (D) Dots show measured exchange energy as a function of distance from the stripe as indicated by the arrow in the inset. The red, blue, and green lines are fits to 1D, 2D, and 3D range functions for indirect exchange. Horizontal error bars are due to the roughness of the Co-ML-stripe edge, whereas the vertical ones are due to the uncertainty in B_{ex}. [67]

curve of a more distant adatom shows a ferromagnetic (FM) coupling (Fig. 1.11B); that is, the adatom magnetization is forced parallel to the stripe at zero field ($J > 0$). An even more distant adatom (Fig. 1.11C) again is antiferromagnetically coupled but with a lower B_{ex} smaller than the stripe coercivity (see arrows) [67].

The interaction energies $J(d)$ determined by the authors from similar magnetization curves of many adatoms are plotted in Fig. 1.11D [67]. A damped oscillatory behavior, which is reminiscent of Ruderman-Kittel-Kasuya-Yosida (RKKY)-like exchange (see Section. 1.4 for detail), is observed [67]. Authors attempt to test whether an RKKY description is appropriate by fitting the data points using range functions $J(d) = J_0 \cdot \cos(2k_F d)/(2k_F d)^n$ with different assumed dimensionalities n (Fig. 1.11D) [67]. A good agreement is found for $n = 1$ and a wavelength of $\lambda_F = 2\pi/k_F \approx 3$ nm [67].

It is interesting to note, that due to mixed dimensionality of the system a fitting with non-integer n might yield a still better agreement. Still the authors have managed to unambiguously show that it is possible to deduce the strength of the exchange coupling of single adatoms from their magnetization curves, a technique which is bound to produce a sharp resonance among the general physics community.

In Chapter 4 we propose an alternative method, which might allow one to probe the exchange coupling in dimers on noble metal surfaces supporting the existence of a surface state.

However, before proceeding to theoretical or experimental investigation of a system or an effect, one has to make sure, that one builds on a steady theoretical fundament. To provide such a fundament for the understanding of the results of this thesis we would like to make a detour into basics of the interaction of impurities. In the next section we will briefly discuss the basics of the chemisorption theory of Anderson, the direct and the indirect interaction of impurities. We will further expand the description onto the general magnetic case and shortly mention the role of special electronic systems (like the surface state) in the interaction mediation.

1.4 Theoretical basics of the interaction of impurities

When foreign nanostructures are introduced into a metallic host the electronic properties of both the host and the impurities are changed. To a large extent the interaction between single impurity atoms (units) is to be held responsible for those changes. If the impurities are situated on (or buried in) a metallic surface the acuteness of the effects can be increased by orders of magnitude. From the physical point of view a surface is a system with reduced dimensionality and thus the interaction is realized simultaneously through both the vacuum and the metallic solid and for this reason it differs significantly from both the interaction between free nanostructures and the interaction between impurities in the bulk of the crystal. Affected by the interaction are the structure of the host, its stability, the vibrational spectra of the adsorbates, migration of particles along the surface. Besides, such characteristics of the surface, as its work function or catalatic activity e.t.c. can be profoundly affected.

Let us agree from the start to classify different types of interaction according to the nature of the virtual quasiparticles that mediate them. Thus the interaction owing to the direct exchange of electrons between adatoms would be called a direct one while the interaction owing to the exchange of electrons through the conduction band would be considered to be indirect. However, this classification of the mechanisms is applicable only as long as we consider the substrate to be a system consisting of noninteracting quasiparticles.

Such interaction mechanisms have been extensively studied in the past. A start was laid by Ruderman and Kittel [69], who considered the interaction of single impurities in solids and were the first to describe the "indirect impurity coupling" mediated by the conduction electrons of a metal. The energy of such a coupling depends on the distance between impurities according to the law [69]

$$E(R) \sim R^{-3} \cdot \cos(2k_F R), \qquad (1.4.1)$$

where R is the separation of impurities and k_F is the Fermi wave vector of host electrons in the corresponding direction. This interaction was shown to play an extremely important role. The long-range oscillating character of it was attributable to the sharp cutoff of the distribution function of conduction electrons in the metal at the Fermi energy, so that the interference of electron waves with the Fermi momentum $k = k_F$ scattered by the impurities leads to the well-known Friedel oscillations of the electron density [70, 71, 72]. As the interaction is mediated by electrons with the energy $\epsilon = \epsilon_F$ the character of the interaction should obviously depend strongly on the form of the Fermi surface (FS) of the metal. Indeed it can be shown (see, e.g. [73]) that if the FS contains cylindrical sections, then in the direction perpendicular to the axis of the cylinder

$$E(R) \sim R^{-2} \cdot \sin(2k_F R), \qquad (1.4.2)$$

and if the FS contains flat sections then in the direction perpendicular to them [74]

$$E(R) \sim R^{-1} \cdot \cos(2k_F R). \qquad (1.4.3)$$

Similar Friedel oscillations also exist in the interaction between atoms adsorbed on the surface of a metal, and for a number of reasons they are stronger in that case. The oscillatory character of the interaction of adatoms was first noted by Grimley [75, 76, 77]. Lau and Kohn [22] as well as Gabovich and Pashitskii [78, 79] noted particularly that for the interaction at the surface to have a long-range character it is sufficient (according to (1.4.2) and (1.4.3)) that the FS of the substrate contains cylindrical or flat sections perpendicular to the surface. In general it can be said, that if the interaction is attributable to the exchange of quasiparticles of the metallic substrate, then it will always have a Friedel oscillating component whose amplitude decreases as some power of R as $R \to \infty$, and the exponent is determined solely by the electronic structure of the substrate.

1.5 Basic ideas from the theory of chemisorption

To understand the particularities of the electronic structure of the adatom-substrate system we will borrow the basic notions of the theory of Newns-Anderson model of chemisorption [28, 27, 80, 81]. We will not elaborate on minor details (as a more detailed discussion can be found elsewhere, e.g. [28, 27]) but rather concentrate on main statements and conclusions. As the basis for the theory serves the Gurney's [82] proposition that when an atom is chemisorbed on a surface an electronic level is formed. It occurs due to the shift of free-adatom electronic levels in the near-surface-field of the metal. If a level happens to occupy the same energy window as the conduction band of the metal, electrons can tunnel from the conduction band into that level of the adatom and vice versa, as a result of which the

1.5 Basic ideas from the theory of chemisorption

discrete electronic level is transformed into a virtual level with a finite half-width δ. This process might be described theoretically by the Hamiltonian

$$H = H_A + H_S + H_{mix}. \tag{1.5.1}$$

To study such a Hamiltonian it is convenient to proceed in a step by step perturbative manner and employ the technique of advanced Green's functions. Then the Hamiltonian term for a single atomic orbital $|A\rangle$ in the second quantization notation would look like $H_A = \epsilon_A c_A^* c_A$ and the Green's function of the free atom is $G_A(\epsilon) = (\epsilon - \epsilon_A - i0)^{-1}$.

The Hamiltonian H_S in (1.5.1) we can assume to be one of a semi-infinite crystal. When describing both the bulk and the surface impurities it is convenient to resort to the use of a point basis $\{|i\rangle\}$, where $|i\rangle$ is a wavefunction localized at the surface of a substrate atom residing at R_j. In this basis the Green's function of the substrate is given by the expression

$$G_S(R_i - R_j, \epsilon) = \langle i|(\epsilon - H_S - i0)^{-1}|j\rangle. \tag{1.5.2}$$

The form of the G_S for special cases of the FS shape will be discussed in detail in section 1.7. It can be shown [28, 27] that at separations R exceeding the lattice constant of the substrate a_0 the Green's function can be expanded using the plane wave formalism as

$$G_S(R, \epsilon) \approx R^{-\nu} g_\nu(\epsilon) \exp(-ik(\epsilon)R), \tag{1.5.3}$$

where $k(\epsilon)$ is the wavevector $(k(\epsilon)||R)$ of a host electron with energy ϵ, ν is a nonnegative integer determined by the structure of the FS of the substrate and $g_\nu(E)$ is a smooth function of energy.

The operator H_{mix} is introduced to account for the coupling of the adatom to the substrate. In the simplest case, when the adatom is bound to a single surface atom $|i\rangle$, the corresponding Hamiltonian term would be given by an expression

$$H_{mix} = V c_A^* c_i + h.c., \tag{1.5.4}$$

where $V = \langle A|H|i\rangle$ is the overlap integral of the wave functions. To account for the coupling to multiple atoms of the substrate a simple procedure can be used which is reduced to a renormalizations of the parameters ϵ_A and V [28, 27, 83, 84].

The perturbative approach to obtaining the total Green's function $\tilde{G} = (\epsilon - H - i0)^{-1}$ of the system, from an non-interacting Green's function G, can be easily realized by means of the Dyson's equation (for more details see 2.2.2 or, e.g., [85, 86, 87]):

$$\tilde{G} = G + G H_{mix} \tilde{G}. \tag{1.5.5}$$

The advantage of this approach is that different terms in the operator H_{mix} can be taken into account systematically. The equation (1.5.5) is a matrix equation, which in the case of a single adsorbate operates with 2×2 matrices (of the $\{|A\rangle, |i\rangle\}$ basis). Knowing the matrix representation of G and considering that $\langle A|G|i\rangle = 0$, the Green's function G_A for a single atomic level can be obtained from (1.5.5):

$$\tilde{G}_A(\epsilon) = \left\langle A \middle| \tilde{G}(\epsilon) \middle| A \right\rangle = \left(\epsilon - \epsilon_A - V^2 G_s(0, \epsilon)\right)^{-1}. \tag{1.5.6}$$

Thus the shift of the electronic level of the adatom accompanying chemisorption equals $\Lambda \equiv \Lambda(\epsilon_A) = \mathfrak{Re}\, V^2 G_S(0, \epsilon_A)$ and its half-width equals $\Delta \equiv \Delta(\epsilon_A) = \mathfrak{Im}\, V^2 G_S(0, \epsilon_A)$. In real systems both the shift and the half-width of the level usually are of the order of 1eV [83].

The density of electronic states of the system is affected by the impurity $\rho(\epsilon) = \pi^{-1} \Im \tilde{G}(\epsilon)$ which allows neighboring impurities to "sense" each other, resulting in what is called an indirect interaction between adatoms. It was first reported by Koutecky [21]. Let us consider two impurities adsorbed on sites $|i\rangle$ and $|j\rangle$ separated by a distance $R = |R_i|$. The dimensionality of the Dyson's equation (1.5.5) is then increased to 2×2 (of the $\{|A\rangle, |B\rangle, |i\rangle, |j\rangle\}$ basis) remaining however easily solvable [88].

The interaction energy of the adatoms can is defined as

$$E_{int}(R) = \langle H(R) \rangle - \langle H(\infty) \rangle. \tag{1.5.7}$$

Details of the representation of $E_{int}(R)$ in terms of the Green's function $\tilde{G}$ are thoroughly discussion in numerous publications (see, e.g. [89, 90, 28, 27] and the references therein). Here, we shall therefore immediately turn to the expression for the indirect-interaction energy, obtained in the framework of the temperature Green's function formalism [85, 91]:

$$E_{ind}(R) = -\tfrac{2}{\pi} \int\limits_{-\infty}^{+\infty} d\epsilon\, f_F(\epsilon)\, \Im\, ln\left(1 - V^4\, \tilde{G}_A^2(\epsilon)\, G_S^2(R, \epsilon)\right),$$
$$f_F(\epsilon) = \{\exp\left[(\epsilon - \epsilon_F)/(k_B T)\right] + 1\}^{-1}. \tag{1.5.8}$$

The factor of two here accounts for the summation over the electron spins. In the $R \to \infty$ limit it can be assumed that $V^2 \tilde{G}_A G_S \ll 1$, which allows the expression (1.5.8) to be simplified:

$$E_{ind}(R) = -\frac{2V^4}{\pi} \Im \int\limits_{-\infty}^{+\infty} d\epsilon\, f_F(\epsilon)\, \tilde{G}_A^2(\epsilon)\, G_S^2(R, \epsilon). \tag{1.5.9}$$

Using the expression (1.5.3) for the Green's function of the substrate and integrating (1.5.9) by parts we obtain in the limit of $R \to \infty$ [28, 27]

$$E_{ind}(R) = -\frac{2V^4}{\pi R^{2\nu+1}} \Re \int\limits_{-\infty}^{+\infty} d\epsilon\, f'_F(\epsilon) \frac{\epsilon \alpha(\epsilon)}{k(\epsilon)}\, \tilde{G}_A^2(\epsilon)\, g_\nu^2(\epsilon)\, \exp(-2ik(\epsilon)R), \tag{1.5.10}$$

where $\alpha^{-1}(\epsilon) = (2\epsilon/k(\epsilon)) \cdot dk(\epsilon)/d\epsilon$. We can further utilize the fact that the localized nature of $f'_F(\epsilon)$ cuts out, upon integration, a narrow region of width $k_B T$ around the Fermi energy. Then, using the expansion $k(\epsilon) = k_F \left[1 + (\epsilon - \epsilon_F)/(2\epsilon_F \alpha_F)\right]$, where $\alpha_F = \alpha(\epsilon)$, and removing the smoothly varying functions from the integrand we obtain the following expression for the indirect interaction energy: [28, 27]

$$E_{ind}(R) \approx \frac{2V^4}{\pi R^{2\nu+1}} \frac{\epsilon_F\, \alpha_F}{k_F} f\left(\frac{R}{R_T}\right) \Re\left[\tilde{G}_A^2(\epsilon_F)\, g_\nu^2(\epsilon_F)\, \exp\left(-i2k_F R\right)\right], \tag{1.5.11}$$

where

$$f(x) = \frac{x}{sh\, x}, \qquad R_T = k_F^{-1} \frac{\alpha_F\, \epsilon_F}{\pi k_B T}. \tag{1.5.12}$$

It is rather natural, that at non-zero temperatures the interactions in the system should decay exponentially with distance in the far zone ($R \gg R_T$):

$$E_{ind}(R) \sim R^{-2\nu} \cos\left(2k_F R + \varphi\right) \exp\left(-\frac{R}{R_T}\right) \tag{1.5.13}$$

The quantity $R_T \sim a_0 \left(\epsilon_F / k_B T\right)$ at room temperatures and a typical values of $E_F \approx 5 eV$ is of the order of hundreds of lattice constants, and thus, at intermediate distances, we can safely neglect the temperature dependence of $E_{ind}(R)$ in our further calculations. There are, however, other factors that affect the R_T. In real crystals is the mean free path of conduction electrons, which is strongly dependant on their scattering by impurities, phonons, etc.

Since at large distances the interaction is mediated by host electrons with energies close to the Fermi energy ϵ_F the period of the oscillations of $E_{ind}(R)$ and the rate of its amplitude's decay are determined solely by the structure of the FS of the substrate along the given vector. Indeed, from (1.5.11) at $T = 0$ we obtain $f(R/R_T) = 1$ and

$$E_{ind}(R) \sim R^{-(2\nu+1)} \cos\left(2k_F R + \varphi\right). \tag{1.5.14}$$

The range of the indirect interaction is longest in the cases of flat ($\nu = 0$) and cylindrical ($\nu = 1/2$) Fermi surfaces, when the interaction is transferred by large groups of electrons. First, however, we shall study the behavior of $E_{ind}(R)$ in the near zone.

1.6 Indirect interaction of adatoms: the near zone

While at large distances $R \gg R_A$ (where R_A is a spatial constant discussed in more detail later) the interaction is transferred mainly by electrons with energies $\epsilon = \epsilon_F$, in the near zone $R \ll R_A$ all electrons in the conduction band of the substrate participate in the interaction; in the process different parts of the FS make different contributions to the interaction energy.

Numerical calculations have shown that the near zone ($R \ll R_A$) is large compared with a_0. For example, in the case of adsorption of hydrogen atoms on the surface of aluminum, in the frame of the "gellium" model, $R_A \approx 5$ Å [92], while for adsorption on the (100) face of a simple cubic lattice R_A equals several lattice constants [93]. The qualitative behavior of the indirect interaction in the near zone can be considered with the help of the expressions (1.5.3) and (1.5.9). Switching to integration over the k space, substituting $d\epsilon = \frac{2\epsilon(k)\alpha(k)}{k} dk$, and $\alpha(k) = \frac{k}{2\epsilon(k)} \cdot \frac{d\epsilon(k)}{dk}$, and removing the smoothly varying function from the integral, we obtain the following expression for the interaction energy: [28, 27]

$$E_{ind}(R) \approx \frac{2V^4}{\pi R^{2\nu}} \frac{2\epsilon_F \alpha_F}{k_F} \, \mathfrak{Im}\left\{g_\nu^2(\epsilon_F) \exp(-i2k_F R) \int^{k_F} dk \tilde{G}_A^2\left(\epsilon(k)\right) \exp\left[-i2(k - k_F)R\right]\right\}. \tag{1.6.1}$$

Regarding the variable k as complex and selecting the lower half-plane for the contour integration [28, 27], using the expression (1.5.6) for $\tilde{G}_A(\epsilon)$ and the expansion

$$\epsilon(k) \approx \epsilon_F \left[1 + (k - k_F)\alpha_F (2k_F)^{-1}\right] \tag{1.6.2}$$

we obtain

$$E_{ind}(R) \approx -\frac{2V^4}{\pi R^{2\nu}} \frac{16 k_F R}{\epsilon_F \alpha_F} \, \mathfrak{Re}\left\{g_\nu^2(\epsilon_F) \left[\exp(-SR) \, \mathrm{Ei}(SR) - (SR)^{-1}\right] \exp\left(-i2k_F R\right)\right\}. \tag{1.6.3}$$

where Ei is the integral exponential function,

$$S = i \frac{4k_F}{\alpha_F} \left(\tilde{\epsilon}_A - \epsilon_F + i\Delta_F\right) \epsilon_F^{-1},$$
$$\tilde{\epsilon}_A = \epsilon_A + \Lambda(\epsilon_F), \quad \Delta_F = \Delta(\epsilon_F). \tag{1.6.4}$$

In the far zone $R \gg R_A$ the asymptotic expression (1.5.14) derives from (1.6.3), and in the near zone $R \ll R_A$ one obtains approximately [28, 27]

$$E_{ind}(R) \approx -\frac{8V^4}{\pi R^{2\nu}} \, \mathfrak{Im} \left[\tilde{G}_A(\epsilon_F) g_\nu^2(\epsilon_F) \exp\left(-i2k_F R\right) \right]. \tag{1.6.5}$$

We note that in the near zone the period of the oscillations is also determined by the Fermi wave vector k_F and does not depend on the parameters of the adatom. [94] The size of the near zone is determined by the parameters of the adatom [28, 27]:

$$R_A = \left|S^{-1}\right| = (4k_F)^{-1} \alpha_F \epsilon_F \left[(\tilde{\epsilon}_A - \epsilon_F)^2 + \Delta_F^2\right]^{-1/2}. \tag{1.6.6}$$

For parameters that are typical for chemisorption systems $|\tilde{\epsilon}_A - \epsilon_F| \approx \Delta_F \approx 0.5$ eV, $\epsilon_F \approx 5$ eV, $k_F \approx a_0^{-1}$ we obtain $R_A \approx 2a_0$, which agrees with the numerical results presented above [92]. If, however, the adatom strongly perturbs the electronic spectrum of the substrate, i.e., if the adatom has a narrow virtual level near the Fermi level ($|\epsilon_A - \epsilon_F| \lesssim \Delta \ll \epsilon_F$), then the near zone is significantly larger ($R \gg a_0$). We emphasize that in this case the amplitude of the indirect interaction is also large [95, 83, 27] and it decays as a function of R in the near zone more slowly than in the far zone [94, 96]. Prior to the appearance of modern calculational facilities and the development of new effective computational techniques, it has been an extremely tough task to calculate the energy E_{ind} in the near zone accurately, since in this case, the electronic structure of the substrate must be taken into account exactly. Calculations performed in a number of papers (see, e.g., Refs. 2, 14, 24, 28, 31, and 34 of [83]) can be regarded as the proof of this statement. Modern numerical methods (like the density functional theory, described in more details in Section 2.1) allow for precise calculations of extremely complex systems which can impossibly be treated purely theoretically. However, simplified models can give us a surprisingly good insight into the general behavior trends of many systems.

1.7 Indirect interaction of adatoms: the asymptotic zone

Since the behavior of the indirect-interaction energy in the far zone $R \gg R_A$ is completely determined by the shape of the FS of the substrate the study of the asymptotic behavior of the function $E_{ind}(R)$ reduces to calculating the Green's function $G_s(R, \epsilon)$. For this it is convenient to employ the strong-coupling approximation, which enables modeling substrates with different electronic structure [94]. Further we will discuss several simple cases which are of interest for the systems discussed in the present thesis.

1.7.1 Flat FS

The simplest model of a metal with a flat FS is a one-dimensional linear chain of atoms with the dispersion law for electrons $\epsilon(k) = \epsilon_c - 2\gamma \, cos(a_0 k)$, where γ is the overlap integral of the orbitals of neighboring atoms, ϵ_c is the center of the conduction band of width W, and $\gamma = W/4$. For convenience we will set, by choosing the system of units, ϵ_c to be 0, $\gamma = 1/2$ and $a_0 = 1$. Then the Green's function of the substrate in the "point" basis ($R = ma_0$)

1.7 Indirect interaction of adatoms: the asymptotic zone

equals [83, 27]

$$\begin{aligned} G_S(m, \epsilon) &= (2\pi)^{-1} \int_{-\pi}^{+\pi} dk \, \exp(imk) \, (\epsilon - \epsilon(k) - i0)^{-1} \\ &= i(-i)^m (1 - \epsilon^2)^{-1/2} \left[\epsilon + i(1 - \epsilon^2)^{1/2} \right]^{|m|} \\ &= \frac{i}{\sin x_0} \exp(-i \, |m| \, x_0), \end{aligned} \quad (1.7.1)$$

where $\cos x_0 = -\epsilon$, $0 \leq x_0 \leq \pi$ [83, 27]. Thus for the interaction of adatoms along closely packed chains of atoms of the substrate (i.e., perpendicularly to the flat sections of the FS) $E_{ind}(R) \sim R^{-1}$ ($\nu = 0$). The situation described corresponds, for example, to the interaction of adatoms adsorbed on a (112) face of a metal and aligned along the [111] direction.

1.7.2 Flat sections of the FS

It is, of course, clear, that the case of a flat section in a FS of a real element is a rather exotic one. However, a Fermi surface of a metal can have flattened sections, or, to be precise, sections, having a large curvature radii. At distances shorter than a certain characteristic length R^*, such patches produce the character of interaction innate to a flat fermi surface [83, 27]. It can be shown [83, 27], that the asymptotic formula $E_{ind}(R) \sim R^{-2}$ is true only at large distances $R \gg R^* = m_v(\epsilon_F) \cdot a_0$. In the near zone ($R \ll R^*$) the preasymptotic formula $E_{ind}(R) \sim R^{-1}$ holds. The parameter R^*, separating the far and the near zones, is proportional to the radius of curvature of the Fermi surface in the selected direction [83, 27].

1.7.3 Open sections of the FS

In some metals, e.g. in fcc stacked crystals like Cu, the almost spherical FS intersects the Brillouin zone boundary, thus leaving open sections. Then in the direction of these sections the interaction can be described by the Green's functions $G_R(R) \sim \exp(-\alpha R)$. So the interaction in these directions is not directly supported by host electrons and decays exponentially [83, 27].

1.7.4 Spherical FS

If the Fermi surface does not display any peculiarities, such as a flattened patch or an open section, the indirect interaction at the surface decays as $E_{ind}(R) \sim R^{-5}$. This behavior, thoroughly discussed, f.e., by Einstein [88] for a simple cubic lattice, significantly differs from case of the interaction in the bulk, which shows cubic decay rate. This can be explained by a much smoother behavior of the density of electronic states at the surface as compared to the bulk [83, 27].

1.7.5 The role of surface states

Some surfaces of close packed crystals support the existence of electronic surface states (SS). In the one-band model, if the perturbation U of the 3D periodic crystal by the surface is taken into account, an additional electronic state arises in the inverted band gap - the Tamm surface state [97]. In a two-band model of a crystal such states arise in the direct gap owing to the mixing of the bands caused by the finiteness of the crystal - Shockley surface state [98]. The fact that such states can also be a carrier for the interatomic interaction was first considered in 1978 by Lau and Kohn [22]. With increasing success of surface probe techniques, such as the STM this topic was reopened both by theoreticians and experimentalists [23, 99]. Hyldgaard and Persson [100], proceeding in a way similar to that, outlined in the present introduction (see also [83, 27]), have examined the indirect interaction mediated by a Shockley surface-state band between adsorbates on the (111) face of noble metal surfaces in the presence of bulk electrons. They stressed the importance of the screening by the finite density of the surrounding conduction-band bulk electrons. They succeeded to obtain a simple non-perturbative description of the adsorbate-induced scattering within the surface-state band. They also managed to provide an analytical estimate of the interaction in the asymptotic case of large $R \gg R^*$ [100]:

$$E_{ind}(R) \approx -\epsilon_F \left(\frac{2\sin(\delta_F)}{\pi}\right)^2 \frac{\sin(2k_F R + 2\delta_F)}{(k_F R)^2}. \tag{1.7.2}$$

The character of the interaction is determined by experimentally accessible parameters: the Fermi-level phase shift, δ_F, which manifests itself in the reference shift of the adsorbate-induced standing-wave LDOS patterns (observed in STM images, e.g. [23]), the Fermi energy ϵ_F of the Shockley surface state band, and the associated $k_\parallel$ Fermi wave vector k_F. This description has proved to be extremely viable and has been corroborated by numerous theoretical and experimental studies [101, 102, 23].

It is, indeed, hard to underestimate the potential scope of possibilities that the surface state mediated interaction presents for nanoscience and nanotechnology. It has already allowed for the discovery of such interesting phenomena as, f.e., the existence of adatom superlattices [103] or the self-assembly of low-dimensional nanostructures [104]. The interaction of single impurities with extended surface structures (steps, stripes, corrals), mediated by surface state electrons, has been shown to have a profound effect on the adsorption properties and the diffusion of single adatoms at the surface [35, 105]. The existence of surface-electrons-stabilized surface lattices has also presented the scientists with an interesting model system for investigating the 2D melting of solids, which was shown to exhibit a peculiar transition from solid to liquid avoiding the intermediate (so called hexatic) phase [106].

1.8 Direct interaction

Last, but not least, let us shed a few words on the direct interaction of adsorbates. When two adatoms (A and B) lie close enough to one another on a substrate for their orbitals Ψ_A and Ψ_B to overlap, the direct mechanism of interaction comes into play. It is fundamentally identical to the interaction leading to the formation of a chemical bond in molecules. In quantum mechanical view of the interaction an overlap between the impurities' orbitals results in the creation of two common orbitals: a bonding $\Psi_+ = \Psi_A + \Psi_B$ and an antibonding $\Psi_- = \Psi_A - \Psi_B$ ones (see Fig. 1.12a). The gain in energy due to the interaction is proportional to the overlap integral $I(R) = \langle \Psi_A | H | \Psi_B \rangle$ and can be attributed to the fact that in the lowest energy state only the energetically favorable orbital Ψ_+ is filled with electrons (see the sketch in Fig. 1.12b). The direct interaction leads to attraction of impurities which falls

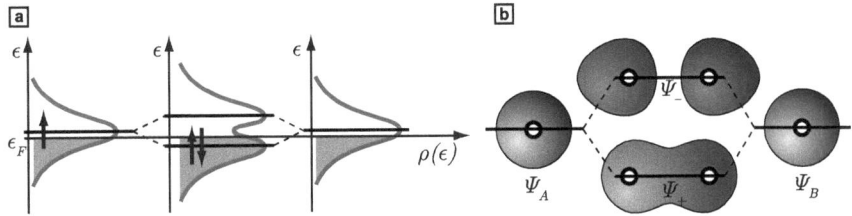

Figure 1.12: a) A schematic representation of the two electronic orbitals formed as a result of the *s*-shells interaction. b) A sketch of the electronic structure modification due to the direct interaction of impurities.

of exponentially with the interatomic separation R following the behavior of the overlap integral: $E(R) \sim I(R) \sim \exp(-\alpha R)$, where α is of the order of the adatoms' outer shell extents [83, 27]. Thus it is obvious that the direct interaction is fundamentally short-ranged and does not affect the processes in the bulk or at the surface for low impurity densities or large interatomic separations.

Other mechanisms of interaction of adatoms (elastic, van-der-Waals [88]) lead to energies that, as a rule, are more than an order of magnitude lower than those discussed above. This means that they can only be significant if the coupling of the impurity to the surrounding host is extremely weak (the so-called physically adsorbtion regime) [83, 27, 88].

1.9 Introducing the magnetism

As the present work is mostly focused on magnetic aspects of the surface and subsurface impurity interaction, it is essential at this point to introduce magnetism into the above presented general framework.

1.9.1 Direct interaction of magnetic impurities

For the direct interaction of impurities it can be relatively easily done as a straightforward extension of Section 1.8 and was indeed done so for the first time by Alexander and Anderson [107]. But instead of replicating their rigorous mathematical derivation we will just briefly summarize the facts complemented by some elementary arguments for the derivation of the Green's function and the corresponding density of states for a ferro- and an antiferromagnetic alignment of impurity magnetic moments (see also the work by Oswald *et al.* [108]). To simplify the derivation we will, without strongly limiting the generality of the results obtained, consider the impurity charges q_1 and q_2 to be equal to the charge of a single noninteracting impurity q. The only parameters left flexible to the effect of the interaction are the magnetic moments of both impurities (m_1 and m_2). This assumption simplifies the self-consistent determination of the moments and the calculation of the interaction energies without changing the results significantly.

In the Anderson model for a single magnetic impurity [81] the impurity level E_0 (more specifically, such a level can be considered to be a virtual bound state, as originally stated by Anderson [81]), becomes spin split due to the exchange interaction: $E = E_0 \pm \frac{1}{2} I \cdot m$,

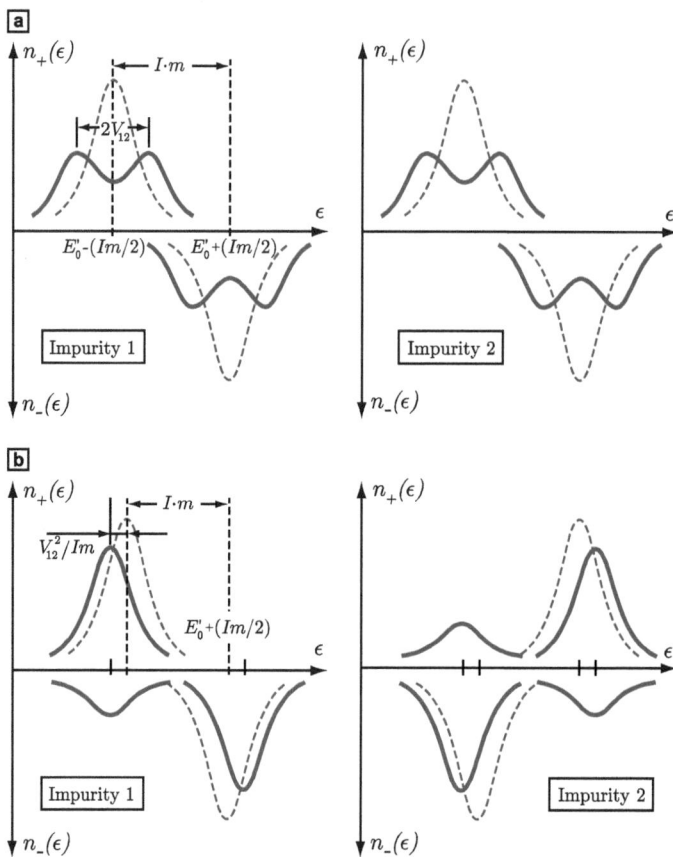

Figure 1.13: Schematic representation of the LDOS for two configurations of a pair of magnetic impurities: (a) ferromagnetic and (b) antiferromagnetic. Dashed curves represent the LDOS of single (noninteracting) impurities in the host. Solid curves represent the LDOS of the pair. The figure has been adopted from [108].

1.9 Introducing the magnetism

where I is the exchange integral and the signs $+$ and $-$ correspond to the spin states $\sigma = -1$ (minority) and $\sigma = +1$ (majority) respectively. In the Anderson model, the embedding of the impurity into a host leads to an energy shift $E_0 \to E_0'$ and to a broadening of the level to a halfwidth of $\Gamma = 2\Delta = 2\pi \langle V_{ih}^2 \rangle n(\epsilon)$ [81, 108], which depends on the impurity-host matrix element for the level in question V_{ih} and its density of states amplitude. The lifetime of such a state is then, obviously, decreased to $\hbar/\Gamma$. The resulting densities of states can then be represented by two Lorentzians with corresponding halfwidths (dashed black curves in Fig. 1.13(a and b)). In most cases, the energy dependence of E_0' and Γ can be safely neglected.

In the Alexander-Anderson model two impurities on neighboring sites interact directly by an overlap matrix element $V_{12} = V$. The impurity moments m_1 and m_2 are parallel in the ferromagnetic configuration ($m_1 = m_2 = m$) and antiparallel in the antiferromagnetic one ($m_1 = -m_2 = m$). The essential point is that due to the interaction V the impurity states with the same spin directions σ hybridise, so that the new eigenfunctions are determined by the equations [108]

$$\left(E_0 - \tfrac{1}{2}\sigma I m_1 - E\right)c_1^\sigma + V c_2^\sigma = 0$$
$$V c_1^\sigma + \left(E_0 - \tfrac{1}{2}\sigma I m_2 - E\right)c_2^\sigma = 0. \quad (1.9.1)$$

For the ferromagnetic case this results in the formation of bonding and antibonding states for each spin direction with a splitting of $2V$:

$$E_\pm = E_0 - \tfrac{1}{2}\sigma I m \pm V,$$
$$c_1^\sigma(\pm) = \pm c_2^\sigma(\pm) = 1/\sqrt{2}. \quad (1.9.2)$$

For the antiferromagnetic pair the local moment is reversed on site 2. Consequently the state $E_0 - \tfrac{1}{2}\sigma I m_1$ at site 1 hybridises with the state $E_0 + \tfrac{1}{2}\sigma I m_2$ at site 2. As a result one obtains the same energy levels for both spin directions, but with strongly differing weights [108]:

$$E_\pm = E_0 \pm W \qquad |c_1^\sigma(\pm)|^2 = \tfrac{1}{2}(1 \pm \sigma\eta) = |c_2^{-\sigma}(\pm)|^2$$
$$W = \left[(\tfrac{1}{2}Im)^2 + V^2\right]^{1/2} \approx \frac{Im}{2} + \frac{V^2}{Im} \qquad \eta = \frac{Im}{2W} \approx 1 - \frac{2V^2}{(Im)^2}. \quad (1.9.3)$$

The embedding of the pair into the host leads again to a change $E_0 \to E_0'$ and a lifetime $\hbar/\Gamma$. By taking this into account the Green's function for the ferromagnetic pair can be expressed as [108]:

$$G_1^\sigma(\epsilon) = G_2^\sigma(\epsilon) = \frac{1}{2}\left(\frac{1}{\epsilon - E_0' + \tfrac{1}{2}\sigma Im - V + i\Gamma} + \frac{1}{\epsilon - E_0' + \tfrac{1}{2}\sigma Im + V + i\Gamma}\right), \quad (1.9.4)$$

and for the antiferromagnetic case it would look like

$$G_1^\sigma(\epsilon) = G_2^{-\sigma}(\epsilon) = \frac{1}{2}\left(\frac{1-\sigma\eta}{\epsilon - E_0' - W + i\Gamma} + \frac{1+\sigma\eta}{\epsilon - E_0' + W + i\Gamma}\right). \quad (1.9.5)$$

The corresponding densities of states, given by the imaginary part of the Green's function, are sketched schematically by solid lines in Fig. 1.13(a and b). For the ferromagnetic case one sees the bonding-antibonding splitting of the single-impurity peaks. In the antiferromagnetic case the single-impurity peaks repel each other and are slightly reduced in intensity. Additional satellite peaks with rather small weights ($\sim 2(V/Im)^2$) appear in the spectra. These are due to tunnelled-through electrons from the neighboring sites. These additional peaks have a tendency to reduce the local moment. Thus, due to the self-consistent renormalization of the local moment, the effective exchange splitting is reduced so that the peaks are backshifted and appear at the same positions as for the isolated impurities [81, 108].

Although at nearest neighbors separations the direct interaction is the most important one, the interaction of the two impurities through the host should not be left completely unmentioned. In more detail the indirect interaction will be discussed in the next section, here we only note, that the hopping of electrons from impurity 1 into the host and from there to impurity 2 leads to an indirect interaction which is described by an energy-dependent matrix element $\lambda_{12}(E)$ superimposed on and competing with the constant direct matrix element $V = V_{12}$ in equations (1.9.4) and (1.9.5) [81].

If we start increasing the separation between the impurities, the relative weight of this RKKY-like interaction will constantly grow.

1.9.2 Indirect interaction of magnetic impurities

It is now taken as a self-evident fact, that the indirect interaction of two magnetic impurities in a paramagnetic host carries a close resemblance to the Ruderman-Kittel-Kasuya-Yosida (RKKY) [109, 69, 110] interaction. It has a spatially oscillating character and can be described in scattering theory terms as an interaction maintained by conduction electrons scattering between two impurity potentials. The propagation of these electrons can be viewed as the propagation of plain waves. Historically, however, [107] it has been derived from a Friedel-Anderson model of localized moments [111, 112], with the use of a Green function formalism and treating the $s-d$ exchange interactions in a Hartree-Fock approximation.

In it's general RKKY form for the exchange interaction energy looks like [113]

$$E_{exc,12} \propto -J_{12}\,\hat{\mathbf{s}}_1 \cdot \hat{\mathbf{s}}_2, \qquad (1.9.6)$$

where $\hat{\mathbf{s}}_{1,2}$ describes a unit vector pointing in the direction of the magnetic moment on one of the impurity sites and J_{12} is an effective exchange parameter. In it's original formulation for magnetic impurities in a metallic-like host the exchange interaction can be implicitly formulated as [113]:

$$E_{exc,12} \propto \frac{\hat{\mathbf{s}}_1 \cdot \hat{\mathbf{s}}_2}{R_{12}^4}\left[2k_F R_{12}\,\cos(2k_F R_{12}) - \sin(2k_F R_{12})\right], \qquad (1.9.7)$$

where R_{12} is the vector connecting the two sites and k_F is the Fermi wave vector of the host. Since its emergence, the formalism has been thoroughly refined and complemented to account for various effects, introduced by host system peculiarities or its relativistic properties. For example, Levi and coworkers [114, 115] have introduced an additional Dzyaloshinsky-Moriya (DM) type [116, 117, 118] term

$$E_{DM} \propto \mathbf{R}_{12} \cdot (\hat{\mathbf{s}}_1 \times \hat{\mathbf{s}}_2). \qquad (1.9.8)$$

to the initial representation of the RKKY interaction to account for the anisotropy field of a spin glass. To derive the term they have considered intermediate spin-orbit scattering by non-magnetic impurities and host atoms (an effective three-site interaction) [113]. Further on,

1.9 Introducing the magnetism

Staunton and coworkers [113] have derived a direct relativistic generalization of the RKKY interaction by considering magnetic anisotropic two-site case. The treated spin polarization and spin-orbit coupling non-perturbatively and on an equal footing, concluding generally, that, apart from being a function of an isotropic spin-spin term $\hat{s}_l \cdot \hat{s}_2$ (see Eq. (1.9.7)), it is also a polynomial function of a Dzyaloshinsky-Moriya term $\mathbf{R}_{12} \cdot (\hat{s}_1 \times \hat{s}_2)$ squared and a pseudo-dipolar term $(\mathbf{R}_{12} \cdot \hat{s}_1)(\mathbf{R}_{12} \cdot \hat{s}_2)$ [113].

In the framework of the present research we will not venture into the range of relativistic effects and anomalous behaviors of the exchange interaction, but will rather endeavor to present a thorough study of various fundamental effects, that can be encountered if one deals with single atomic-sized magnetic impurities in a metallic host. In such a case, we can consider a direct extensions of the ideas contained in sections. 1.6 and 1.7 onto the case of magnetic impurities. As has been shown in the above mentioned sections (see, e.g., Section 1.5), the indirect interaction in the nonmagnetic case can be represented, in the matrix element V_i^σ notation, as

$$E_{ind}^{NM} \sim \Im\mathfrak{m} \int d\epsilon \, f_F(\epsilon) \mathrm{Tr}\left\{ (V_1^+ V_1^-) G_{12} (V_2^+ V_2^-) G_{21} \right\}. \tag{1.9.9}$$

If we now consider, that the exchange parameter J_{12} can be represented it terms of the same parameters as

$$J_{12} \sim \Im\mathfrak{m} \int d\epsilon \, f_F(\epsilon) \mathrm{Tr}\left\{ (V_1^- V_1^-) G_{12} (V_2^- V_2^-) G_{21} \right\}, \tag{1.9.10}$$

it becomes clear, that the general trends for the E_{ind} behavior established in sections 1.6 and 1.7 can be safely transferred onto the indirect magnetic interaction of single impurities. We thus can summarize that we have obtained the same dependencies for the indirect exchange interaction of magnetic impurities as were presented at the beginning of the chapter. The exchange interaction in a 3-dimensional (3D) system follows the law [113]

$$E_{exc}(R) \sim R^{-3} \cdot \cos(2k_F R), \tag{1.9.11}$$

where $R = R_{12}$ is the separation of the two impurities. In the 2D case the interaction can be presented as [113]

$$E_{exc}(R) \sim R^{-2} \cdot \sin(2k_F R), \tag{1.9.12}$$

and in the 1D case (e.g. if the FS contains flat sections) [113]

$$E_{exc}(R) \sim R^{-1} \cdot \cos(2k_F R). \tag{1.9.13}$$

Such simple approximation often provide an accurate description of the realistic physical picture. As an example, the famous phenomenon of the interlayer exchange coupling can be considered [119, 120, 121, 122, 123, 12, 11]. The long history of this phenomenon's investigation has shown, that by refining the theoretical model one can achieve a striking agreement with the experimental observations. Yet it is a very taxing task, as, even for a simple metallic system, a general case analytical derivation of J_{12} might prove to be a nigh on impossible procedure. That is why we employ the general theoretical estimations for identifying potentially interesting systems and as a reference point for further *ab initio* calculations.

Main aims of the study

Although the amassed experimental expertise and simple theoretical models provided in the introductory chapters give one a good estimate of a possible behavior of a nanoscale system, they are usually too crude to provide an accurate description of finer effects and are thus often useless in studying new effects and testing new theories. In the present study we rely on an *ab initio* KKR Green's function method to investigate magnetic properties of subnanoscale units at surfaces to clarify and generalize the following points:

- Coupling of single magnetic adatoms to a buried magnetic monolayer: single spin manipulation.
- Effect of a buried magnetic monolayer on the interatomic exchange coupling in surface addimers.
- Tailoring the exchange interaction of single adatoms by linking them with paramagnetic chains.
- Utilizing the quantum confinement of surface state electrons on selfassembled structures to tune the exchange coupling of single magnetic adatoms adsorbed on top of them.
- Surface state localization at magnetic addimers as a probe for the interatomic exchange interaction.
- Possibility to study the magnetism of buried structures by analyzing the polarization of surface electrons.
- Possibility to determine the coupling of buried structures to each other.

Chapter 2

Theoretical approach used in the study

The most general approach possible for the descriprion of the electronic properties of a solid consisting of N electrons and M nuclei is the solution of a $3(N+M)$-dimensional Srödinger equation for the many-body wave-function $\psi\left(\mathbf{r}_{1}, \mathbf{r}_{2}, \ldots, \mathbf{r}_{N}, \mathbf{R}_{1}, \mathbf{R}_{2}, \ldots, \mathbf{R}_{M}\right)$ where $\mathbf{r}_{n}$ and $\mathbf{R}_{m}$ are the coordinates of single electrons and nuclei respectively [124, 125]. Even assuming the *adiabatic Born and Oppenheimer approximation* we are left with a $3N$-dimensional Schrödinger equation which is quite impossible to tackle with modern (or ever conceivable) theoretical or calculational methods. In the present chapter the Kohn-Sham density functional theory (DFT), allowing for an effective calculational treatment of such many-electron problems, is discussed and the Korringa-Kohn-Rostoker (KKR) Green's function (GF) method is introduced which can be utilized to obtain a solution of the resulting single-particle Kohn-Sham problem.

2.1 Density functional theory

The Hohenberg-Kohn density functional theory (being an exactification of both the Thomas-Fermi and the Hartree theories) was developed by Kohn and Hohenberg around 1964 [126]. It recasts the problem of the solution of a many-electron Schrödinger equation in terms of the electronic density distribution $n(\mathbf{r})$ and a universal functional of the density $E_{xc}\left[n(\mathbf{r})\right]$. Thus is becomes possible to replace the necessarily approximate solution of a many-body Schrödinger equation by a problem of finding adequate approximations to the exchange functional and then solving a single-particle electronic equation.

2.1.1 Hohenberg - Kohn theorems

The base for the theoretical justification of the DFT are two theorems proving the possibility of an above-mentioned treatment of a many-body system [125, 127, 128, 129]. If we consider a system of N interacting electrons in a non-degenerate ground state ψ represented by an external potential $v(\mathbf{r})$, then the theorem states, that [126]

Theorem 1 *(Also called the original Lemma) The ground state external potential $v(\mathbf{r})$, and hence the total energy, is uniquely determined by the electron density $n(\mathbf{r})$ (to within an additive constant).*

Though the original formulation contains the limitation of the non-degeneracy of the ground state and was conceived for non-relativistic potentials and spin-less particles, it has bin long since purged of its restriction and considerably generalized.

Theorem 2 *(Also called the Variational Theorem) There exists a functional $F[n'(\mathbf{r})]$ defined for all non-degenerate ground state densities $n'(\mathbf{r})$ such that, for a given $v(\mathbf{r})$, the quantity*

$$E_{v(\mathbf{r})}[n'(\mathbf{r})] = \int v(\mathbf{r})\, n'(\mathbf{r})\, d\mathbf{r} + F[n'(\mathbf{r})] \qquad (2.1.1)$$

has its unique minimum for the correct ground-state density, $n'(\mathbf{r}) = n(\mathbf{r})$, associated with $v(\mathbf{r})$ [126].

The functional $F[n'(\mathbf{r})]$ can be represented as

$$F[n'(\mathbf{r})] \equiv \left(\psi_{n'(\mathbf{r})}, (T_S + U)\psi_{n'(\mathbf{r})}\right), \qquad (2.1.2)$$

where $\psi_{n'(\mathbf{r})}$ is the ground state associated with $n'(\mathbf{r})$, and T_S and U are the kinetic energy and the Coulomb repulsion operators. In terms of the ground state density this functional can be expressed as

$$F[n(\mathbf{r})] \equiv T_S[n(\mathbf{r})] + \frac{1}{2}\iint \frac{n(\mathbf{r})n(\mathbf{r}')}{|\mathbf{r}-\mathbf{r}'|}\, d\mathbf{r}d\mathbf{r}' + E_{xc}[n(\mathbf{r})] \qquad (2.1.3)$$

It can be noted, that if the last term $E_{xc}[n(\mathbf{r})]$ in equation (2.1.1) is neglected equations (2.1.1) and (2.1.3) turn into the set of Hartree equations. Hence, all the explicit many-body effect beyond the Hartree mean field approximation are encapsulated in the exchange correlation functional $E_{xc}[n]$. Thus it can be said, that the ground state energies and corresponding densities can be obtained by minimizing (2.1.1) with respect to $n[\mathbf{r}]$.

2.1.2 Kohn-Sham (KS) equations

For an efficient minimization of the energy functional (2.1.1) it is convenient to reformulate the Hohenberg Kohn variational principle in form of exact single particle self consistent equations, similar to the Hartree ones. It has been done by Kohn and Sham [130] within an assumption that the total number of electrons N is fixed (i.e. $\int n(\mathbf{r})d\mathbf{r} = N$). This constraint was introducing by the Lagrange multiplier, μ, chosen so that

$$\frac{\delta}{\delta n(\mathbf{r})}\left[E[n(\mathbf{r})] - \mu N\right] = 0, \qquad (2.1.4)$$

which yields the Euler-Lagrange equation

$$\frac{\delta E[n(\mathbf{r})]}{\delta n(\mathbf{r})} = \mu, \qquad (2.1.5)$$

which can be rewritten, considering (2.1.1) and (2.1.3), as

$$\frac{\delta T_s[n(\mathbf{r})]}{\delta n(\mathbf{r})} + v_{eff}(\mathbf{r}) = \mu, \qquad (2.1.6)$$

2.1 Density functional theory

where $v_{eff}(\mathbf{r})$ is an effective single-particle potential:

$$v_{eff}(\mathbf{r}) = v(\mathbf{r}) + \int \frac{n(\mathbf{r}')}{|\mathbf{r}-\mathbf{r}'|} d\mathbf{r}' + v_{xc}(\mathbf{r}), \qquad (2.1.7)$$

accounting for the many body effects by the functional derivative

$$v_{xc}[\mathbf{r}] = \frac{\delta E_{xc}[n(\mathbf{r})]}{\delta n(\mathbf{r})}. \qquad (2.1.8)$$

Now, if one considers a system of non-interacting electrons moving in an external potential $v_{xc}[\mathbf{r}]$, as defined in (2.1.8), one could easily find the ground state energy and density by solving a one-electron Schrödinger-like equation, which carries the name of Konh and Sham:

$$\left\{-\frac{\hbar^2}{2m} + v(\mathbf{r}) + \int \frac{n(\mathbf{r}')}{|\mathbf{r}-\mathbf{r}'|} d\mathbf{r}' + v_{xc}(\mathbf{r})\right\} \psi_k(\mathbf{r}) = \epsilon_k \psi_k(\mathbf{r}), \qquad (2.1.9)$$

where the density is defined by the wave function $\psi_k(\mathbf{r})$ as

$$n(\mathbf{r}) = \sum_{k=1}^{N} |\psi_k(\mathbf{r})|^2. \qquad (2.1.10)$$

It must be noted here, that single particle wave function $\psi_k(\mathbf{r})$ alone only describe abstract quasiparticles without an actual physical meaning, so only a self consistent solution of (2.1.9) and (2.1.10) would yield a description of a real physical system, which is described by the overall electronic density $n(\mathbf{r})$.

2.1.3 The local density approximation

As it has been mentioned before, one of the major challenges in the DFT is finding adequate approximations to the exchange correlation functional. And formal DFT might have not evolved in what it is now, had there not been a simple and practical approximation for E_{xc}, the local density approximation (LDA), which has yielded surprisingly accurate results for a wide range of systems:

$$E_{xc}^{LDA}[n(\mathbf{r})] \equiv \int n(\mathbf{r}) \epsilon_{xc}(n(\mathbf{r})) \, d\mathbf{r}, \qquad (2.1.11)$$

where $\epsilon_{xc}(n)$ is a very accurately known exchange correlation energy per particle of a *uniform* electron gas of density n. The exchange correlation potential can thus be expressed as:

$$v_{xc}[\mathbf{r}] = \frac{\delta}{\delta n(\mathbf{r})} [\epsilon_{xc}(n)n(\mathbf{r})]. \qquad (2.1.12)$$

Thus the problem of the exchange correlation in an inhomogeneous system is replaced by calculating the $\epsilon_{xc}(n)$ for a homogeneous electron gas [127, 128, 125, 129]. Though quite accurate for many different systems, the LDA is assumed to be at it's best for systems with a slowly varying electron density.

2.2 Green's function method

Having established a general procedure of solving, in a rather simple manner, many-body quantum problems DFT itself does not give an answer as to how one could obtain the actual numerical solution of the resulting Kohn-Sham equation (2.1.9) for an arbitrary system. Having selected an appropriate exchange correlation functional, encapsulating all the necessary many-body effects, one has to obtain self consistent wave functions of the system.

The most straightforward way to do this is to try to expand them in a certain basis set ϕ_i^{basis} and solve the resulting secular equation to obtain the expansion coefficients. Several effective methods have been devised utilizing this approach mainly differing in the expansion set of choice. They include the linear combination of atomic orbitals (LCAO), pseudopotential technique, the (linear) augmented plane wave method ((L)APW) and several others [124, 125].

A different, more subtle, approach does not deal with effective quasi-particle wave functions, but rather exploits the properties of the Green's function of a Kohn-Sham Hamiltonian to obtain the electronic density of the system. This technique has been devised by Korringa [131] and adapted by Kohn and Rostoker [132] thus acquiring the name of Korringa-Kohn-Rostoker (KKR) Green's function method. It allows one to describe electronic properties of large complex systems bulk, interface or localized structure character.

In present section some basic notions of the KKR Green's function method are presented describing the interrelation of the Green's function with the system's electronic properties, followed by a discussion of a possibility to express the Green's function of a complex system in terms of perturbation theory. Further on some elements of the multiple scattering theory are introduced. Concluding the section the general workflow of the KKR method is presented.

2.2.1 Green's functions: definition and basic properties

A resolvent of a Hermitian operator (in our case the Kohn-Sham Hamiltonian) is defined as follows

$$\hat{\mathcal{G}}(z) = \left(z\hat{\mathcal{I}} - \hat{\mathcal{H}}\right)^{-1} \quad , \quad z = \epsilon + i\delta \quad , \quad \hat{\mathcal{G}}(z^*) = \hat{\mathcal{G}}(z)^\dagger, \qquad (2.2.1)$$

where $\hat{\mathcal{I}}$ is the unity operator. Any representation of such a resolvent is called a Green's function. For our purposes the configuration space representation is the most logical choice:

$$\left\langle \mathbf{r} \middle| \hat{\mathcal{G}}(z) \middle| \mathbf{r}' \right\rangle = G(\mathbf{r}, \mathbf{r}'; z). \qquad (2.2.2)$$

Assuming that the side limits of $\hat{\mathcal{G}}(z)$ are defined by

$$\lim_{|\delta| \to 0} \hat{\mathcal{G}}(z) = \begin{cases} \hat{\mathcal{G}}^+(\epsilon), & \text{if } \delta > 0 \\ \hat{\mathcal{G}}^-(\epsilon), & \text{if } \delta < 0 \end{cases}, \qquad (2.2.3)$$

$$\hat{\mathcal{G}}^+(\epsilon) = \hat{\mathcal{G}}^-(\epsilon)^\dagger, \qquad (2.2.4)$$

we come to the following property of $\hat{\mathcal{G}}(z)$:

$$\mathfrak{Im}\hat{\mathcal{G}}^+(\epsilon) = \frac{1}{2i}\left(\hat{\mathcal{G}}^+(\epsilon) - \hat{\mathcal{G}}^-(\epsilon)\right), \qquad (2.2.5)$$

2.2 Green's function method

or, by making use of the properties of the Dirac delta functions,

$$\mathfrak{Im}\mathrm{Tr}\hat{\mathcal{G}}^{\pm}(\epsilon) = \mp \sum_{k} \delta\left(\epsilon - \epsilon_{k}\right), \tag{2.2.6}$$

$$n(\epsilon) = \mp \frac{1}{\pi} \mathfrak{Im}\mathrm{Tr}\hat{\mathcal{G}}^{\pm}(\epsilon), \tag{2.2.7}$$

where Tr is the trace of an operator and $n(\epsilon)$ is the density of states of the Kohn-Sham Hamiltonian with a discrete eigenvalue spectrum $\{\epsilon_k\}$. Let us agree, for the sake of convenience and without limiting the generality, to use only the $\hat{\mathcal{G}}^+$ side limit of the Green's function in future mathematical layouts layouts. As can be seen, the generalization to the case of $\hat{\mathcal{G}}^{\pm}$ can be easily achieved by correct choice of the signs.

In space-resolved energy representation, for $\hat{\mathcal{G}}^+$ (an outgoing wave at **r**) the same dependence can be represented as

$$n(\mathbf{r}, \epsilon) = -\frac{1}{\pi} \mathfrak{Im} G(\mathbf{r}, \mathbf{r}, \epsilon), \tag{2.2.8}$$

which leads us either to the expression for the spectral density of states:

$$n(\epsilon) = -\frac{1}{\pi} \mathfrak{Im} \int G(\mathbf{r}, \mathbf{r}, \epsilon)\, \mathbf{d}^3\mathbf{r}, \tag{2.2.9}$$

or, by integration of the Green's function over the energy up to the Fermi level ϵ_F, to the expression for the corresponding charge density:

$$\rho(\mathbf{r}) = -\frac{1}{\pi} \mathfrak{Im} \int_{-\infty}^{\epsilon_F} G(\mathbf{r}, \mathbf{r}, \epsilon)\, \mathbf{d}\epsilon. \tag{2.2.10}$$

Thus, if we find a way to compute the Green's function, we would get access to all relevant physical properties of the system.

2.2.2 The Dyson equation

Dealing with realistic solid state systems it is often convenient [87] to resort to perturbative approaches. Thus it might be useful to represent the Hamiltonian $\hat{\mathcal{H}}$ of our system as a sum of unperturbed term $\hat{\mathcal{H}}_0$ and a (Hermitian) perturbation $\hat{\mathcal{V}}$:

$$\hat{\mathcal{H}} = \hat{\mathcal{H}}_0 + \hat{\mathcal{V}}. \tag{2.2.11}$$

This is a convenient way to describe, for instance, the scattering of electrons on a single perturbing potential in an otherwise ideal system as the corresponding resolvents $\hat{\mathcal{G}}_0(z)$ and $\hat{\mathcal{G}}(z)$ can be coupled in terms of the Dyson equation,

$$\begin{aligned}\hat{\mathcal{G}}(z) &= \hat{\mathcal{G}}_0(z) + \hat{\mathcal{G}}(z)\, \hat{\mathcal{V}}\, \hat{\mathcal{G}}_0(z) \\ &= \hat{\mathcal{G}}_0(z) + \hat{\mathcal{G}}_0(z)\hat{\mathcal{V}}\, \hat{\mathcal{G}}(z),\end{aligned} \tag{2.2.12}$$

which can be expanded iteratively into a Born series:

$$\begin{aligned}\hat{\mathcal{G}}(z) &= \hat{\mathcal{G}}_0(z) + \hat{\mathcal{G}}_0(z)\hat{\mathcal{V}}\hat{\mathcal{G}}_0(z) + \hat{\mathcal{G}}_0(z)\hat{\mathcal{V}}\hat{\mathcal{G}}_0(z)\hat{\mathcal{V}}\hat{\mathcal{G}}_0(z) + \ldots \\ &= \hat{\mathcal{G}}_0(z) + \hat{\mathcal{G}}_0(z)\left[\hat{\mathcal{V}} + \hat{\mathcal{V}}\hat{\mathcal{G}}_0(z)\hat{\mathcal{V}} + \hat{\mathcal{V}}\hat{\mathcal{G}}_0(z)\hat{\mathcal{V}}\hat{\mathcal{G}}_0(z)\hat{\mathcal{V}} + \ldots\right]\hat{\mathcal{G}}_0(z).\end{aligned} \tag{2.2.13}$$

The term in square brackets is usually called the T-operator

$$\hat{\mathcal{T}} = \hat{\mathcal{V}} + \hat{\mathcal{V}}\hat{\mathcal{G}}_0(z)\hat{\mathcal{V}} + \hat{\mathcal{V}}\hat{\mathcal{G}}_0(z)\hat{\mathcal{V}}\hat{\mathcal{G}}_0(z)\hat{\mathcal{V}} + \ldots \qquad (2.2.14)$$

Thus the Dyson equation can be reformulated as follows:

$$\hat{\mathcal{G}}(z) = \hat{\mathcal{G}}_0(z) + \hat{\mathcal{G}}_0(z)\hat{\mathcal{T}}\hat{\mathcal{G}}_0(z) \qquad (2.2.15)$$

2.2.3 The Lloyd formula

If we reformulate (2.2.7), substituting (2.2.15) into it, we obtain

$$\begin{aligned} n(\epsilon) &= -\frac{1}{\pi}\Im\mathrm{Tr}\left(\hat{\mathcal{G}}_0^+(\epsilon) + \hat{\mathcal{G}}_0^+(\epsilon)\hat{\mathcal{T}}^+(\epsilon)\hat{\mathcal{G}}_0^+(\epsilon)\right) \\ &= n_0(\epsilon) + \delta n(\epsilon), \end{aligned} \qquad (2.2.16)$$

where

$$n_0(\epsilon) = -\frac{1}{\pi}\Im\mathrm{Tr}\left(\hat{\mathcal{G}}_0^+(\epsilon)\right), \qquad (2.2.17)$$

$$\begin{aligned} \delta n(\epsilon) &= -\frac{1}{\pi}\Im\mathrm{Tr}\left(\hat{\mathcal{G}}_0^+(\epsilon)\hat{\mathcal{T}}^+(\epsilon)\hat{\mathcal{G}}_0^+(\epsilon)\right) \\ &= -\frac{1}{\pi}\Im\mathrm{Tr}\left(\hat{\mathcal{G}}_0^+(\epsilon)^2\hat{\mathcal{T}}^+(\epsilon)\right), \end{aligned} \qquad (2.2.18)$$

and, making use of the identity

$$\frac{\mathrm{d}\hat{\mathcal{G}}(z)}{\mathrm{d}z} = -\hat{\mathcal{G}}(z)^2, \qquad (2.2.19)$$

$$\delta n(\epsilon) = \frac{1}{\pi}\Im\mathrm{Tr}\left(\frac{\mathrm{d}\hat{\mathcal{G}}_0^+(\epsilon)}{\mathrm{d}\epsilon}\hat{\mathcal{T}}^+(\epsilon)\right). \qquad (2.2.20)$$

Based on the definition of $\hat{\mathcal{T}}$ one further can derive that

$$\frac{\mathrm{d}\hat{\mathcal{T}}(z)}{\mathrm{d}z} = \hat{\mathcal{T}}(z)\frac{\mathrm{d}\hat{\mathcal{G}}_0(z)}{\mathrm{d}z}\hat{\mathcal{T}}(z), \qquad (2.2.21)$$

and therefore

$$\hat{\mathcal{T}}(z)^{-1}\frac{\mathrm{d}\hat{\mathcal{T}}(z)}{\mathrm{d}z} = \frac{\mathrm{d}\hat{\mathcal{G}}_0(z)}{\mathrm{d}z}\hat{\mathcal{T}}(z), \qquad (2.2.22)$$

which can be substituted into (2.2.20) to yield

$$\begin{aligned} \delta n(\epsilon) &= -\frac{1}{\pi}\Im\mathrm{Tr}\left(\hat{\mathcal{T}}^+(\epsilon)\frac{\mathrm{d}\hat{\mathcal{T}}^+(\epsilon)}{\mathrm{d}\epsilon}\right) = \\ &= -\frac{\mathrm{d}}{\mathrm{d}\epsilon}\left(\frac{1}{\pi}\Im\mathrm{Tr}\ln\hat{\mathcal{T}}^+(\epsilon)\right). \end{aligned} \qquad (2.2.23)$$

The integrated DOS can then be directly expressed as

$$\hat{N}(\epsilon) = N_0(\epsilon) + \delta N(\epsilon), \qquad (2.2.24)$$

2.2 Green's function method

where

$$N_0(\epsilon) = \int_{-\infty}^{\epsilon} n_0(\epsilon')\, \mathrm{d}\epsilon' \qquad (2.2.25)$$

and

$$\delta N(\epsilon) = \frac{1}{\pi} \mathfrak{Im}\,\mathrm{Tr}\ln \hat{\mathcal{T}}^+(\epsilon) \qquad (2.2.26)$$

The above expression is usually referred to as the *Lloyd formula* [129, 133, 134, 135].

2.2.4 Lippmann-Schwinger equation

Before we proceed to the description of the electron scattering by various perturbing potentials we have to address the problem of obtaining the wave functions for the new state of the perturbed system. Suppose the generalized wave functions of unperturbed $\hat{\mathcal{H}}_0$ and perturbed $\hat{\mathcal{H}} = \hat{\mathcal{H}}_0 + \hat{\mathcal{V}}$ systems are $\varphi_i(\epsilon)$ and $\psi_i(\epsilon)$, respectively. Then

$$\left(\epsilon\hat{\mathcal{I}} - \hat{\mathcal{H}}_0\right)\varphi_i(\epsilon) = 0$$
$$\left(\epsilon\hat{\mathcal{I}} - \hat{\mathcal{H}}_0\right)\psi_i(\epsilon) = \hat{\mathcal{V}}\psi_i , \qquad (2.2.27)$$

which leads, under an Ansatz that $\psi_i(\epsilon) = \varphi_i(\epsilon) + \delta\psi_i(\epsilon)$, to

$$\left(\epsilon\hat{\mathcal{I}} - \hat{\mathcal{H}}_0 - \hat{\mathcal{V}}\right)\delta\psi_i = \hat{\mathcal{V}}\varphi_i. \qquad (2.2.28)$$

Applying to $\epsilon\hat{\mathcal{I}} - \hat{\mathcal{H}}$ the definition of the Green's function (2.2.1):

$$\psi_i(\epsilon) = \varphi_i(\epsilon) + \hat{\mathcal{G}}(\epsilon)\hat{\mathcal{V}}\varphi_i(\epsilon), \qquad (2.2.29)$$

or in terms of the T-operator (2.2.14):

$$\psi_i(\epsilon) = \varphi_i(\epsilon) + \hat{\mathcal{G}}_0(\epsilon)\hat{\mathcal{T}}\varphi_i(\epsilon). \qquad (2.2.30)$$

Considering the side limits of the $\hat{\mathcal{G}}$ and the $\hat{\mathcal{T}}$ operators it must be noted that two different solutions $\psi_i^\pm(\epsilon)$ exist. Either of the latter equation (2.2.29), (2.2.30) is called the *Lippmann-Schwinger equation* [129, 133, 134, 135], which relates the generalized wave functions of the perturbed system to those of the unperturbed one.

2.2.5 Single- and multiple-site T-operator

If our system has a single perturbing potential $\hat{\mathcal{V}}_i$. Then the corresponding T-operator is called a *single-site T-operator* and is usually denoted by $\hat{t}_i(z)$ or just $\hat{t}_i$:

$$\hat{t}_i = \hat{\mathcal{V}}_i + \hat{\mathcal{V}}_i\hat{\mathcal{G}}_0\hat{t}_i = \left(\hat{\mathcal{I}} - \hat{\mathcal{V}}_i\hat{\mathcal{G}}_0\right)^{-1}\hat{\mathcal{V}}_i \qquad (2.2.31)$$

In a more general case of a system with an ensemble of of N scatterers we can operate with an effective single particle potential $V(\mathbf{r}) = \left\langle \mathbf{r}\left|\hat{\mathcal{V}}\right|\mathbf{r}\right\rangle$ which can be represented as a sum of individual (effective) potentials

$$V(\mathbf{r}) = \sum_{i=1}^{N} V_i(\mathbf{r}) \qquad (2.2.32)$$

defined in corresponding spatial domains D_{V_i} disjoint in $\mathbb{R}^3$:

$$D_{V_i} \cap D_{V_j} = \delta_{ij} D_{V_i}$$

$$V_i(\mathbf{r}) = 0 \,, \; \mathbf{r} \notin D_{V_i}$$

The T-operator for a system with such a potential can be expressed in terms of single-site T-operators t_i for individual potentials V_i, as

$$\begin{aligned}\hat{\mathcal{T}}_{ms} &= \sum_i t_i + \\ &+ \sum_{i,j} t_i \left(1 - \delta_{ij}\right) t_j + \\ &+ \sum_{i,j,k} t_i \left(1 - \delta_{ij}\right) t_j \left(1 - \delta_{jk}\right) t_k + \\ &+ \sum_{i,j,k,l} t_i \left(1 - \delta_{ij}\right) t_j \left(1 - \delta_{jk}\right) t_j \left(1 - \delta_{kl}\right) t_l + \\ &+ \ldots \; .\end{aligned} \qquad (2.2.33)$$

The first term of the sum series accounts for all the single scattering processes in the system, the second one - for the double scattering, the third - for the triple and so on. Consequently, the Green's function of the system and thus it's electronic properties are determined by the multiplicity of all possible scattering sequences.

2.2.6 Structural Green's function

A different way of summating over multiple scatterers is in terms of the so-called *scattering path operators* (SPO):

$$\begin{aligned}\hat{\tau}^{ij} &= \hat{t}_i \delta_{ij} + \\ &+ \hat{t}_i \left(1 - \delta_{ij}\right) \hat{t}_j + \\ &+ \sum_m \hat{t}_i \left(1 - \delta_{im}\right) \hat{t}_m \left(1 - \delta_{mj}\right) \hat{t}_j + \\ &+ \sum_{m,n} \hat{t}_i \left(1 - \delta_{im}\right) \hat{t}_m \left(1 - \delta_{mn}\right) \hat{t}_n \left(1 - \delta_{nj}\right) \hat{t}_j + \\ &+ \ldots \; .\end{aligned} \qquad (2.2.34)$$

The multiple-site T-operator can then be expressed as

$$\hat{\mathcal{T}}_{ms} = \sum_{i,j} \hat{\tau}^{ij} \,, \qquad (2.2.35)$$

Then the multiple scattering equation for $\hat{\mathcal{G}}$ can be written as

$$\hat{\mathcal{G}} = \hat{\mathcal{G}}_0 + \sum_{i,j} \hat{\mathcal{G}}_0 \, \hat{\tau}^{ij} \, \hat{\mathcal{G}}_0 \,. \qquad (2.2.36)$$

2.2 Green's function method

This brings us to the following Dyson equation

$$\begin{aligned}\hat{\tau}^{ij} &= \hat{t}_i\,\delta_{ij} + \sum_m t_i\,\hat{\mathcal{G}}_0(1-\delta_i m)\,\hat{\tau}^{mj} \\ &= \hat{t}_i\,\delta_{ij} + \sum_m \hat{\tau}^{im}\,\hat{\mathcal{G}}_0(1-\delta_m j)\,\hat{t}_j,\end{aligned} \qquad (2.2.37)$$

which can be reformulated in terms of the so-called *structural resolvent* or *structural Green's function* $\hat{\mathcal{G}}^{ij}$:

$$\hat{\mathcal{G}}^{ij} = \hat{\mathcal{G}}_0(1-\delta_i j) + \sum_{n,m}\hat{\mathcal{G}}_0(1-\delta_{im})\hat{\tau}^{mn}\hat{\mathcal{G}}_0(1-\delta_{nj}), \qquad (2.2.38)$$

as

$$\hat{\tau}^{ij} = \hat{t}_i\,\delta_{ij} + \hat{t}_i\,\hat{\mathcal{G}}^{ij}\,\hat{t}_j. \qquad (2.2.39)$$

Combining equations (2.2.36) and (2.2.39) we can now represent the Green's function of a system with multiple perturbing potentials as

$$\hat{\mathcal{G}} = \hat{\mathcal{G}}^i\delta_{ij} + \sum_{i,j} \hat{\mathcal{G}}_0\,\hat{t}_i\,\hat{\mathcal{G}}^{ij}\,\hat{t}_j\,\hat{\mathcal{G}}_0, \qquad (2.2.40)$$

where $\hat{\mathcal{G}}^i$ are the resolvents of single-site scattering problems for respective domains D_{V_i}. The use name "structural" becomes clear if we look at the structure of relation (2.2.40). In such a representation the single-site Green's functions of single domains are separated and independent of their geometrical arrangement, which is accounted for by the structural Green's function $\hat{\mathcal{G}}^{ij}$ [129, 133, 134, 135].

2.2.7 Fundamental KKR equation

For purposes of numerical calculations it is most convenient to reformulate (2.2.37) and (2.2.38) in terms of supermatrices. For the following set of supermatrices [129, 134, 136]:

$$\begin{aligned}\hat{\mathbf{G}} &= \left\{\hat{\mathcal{G}}^{ij}\right\}, \\ \hat{\mathbf{t}} &= \left\{\hat{t}_i\delta_{ij}\right\}, \\ \hat{\mathbf{G}}_0 &= \left\{\hat{\mathcal{G}}_0\left(1-\delta_{ij}\right)\right\}, \\ \hat{\boldsymbol{\tau}} &= \left\{\hat{\tau}^{ij}\right\},\end{aligned} \qquad (2.2.41)$$

equations (2.2.37) and (2.2.38) can be rewritten as

$$\hat{\boldsymbol{\tau}} = \left(\hat{\mathbf{t}}^{-1} - \hat{\mathbf{G}}_0\right)^{-1}, \qquad (2.2.42)$$

$$\hat{\mathbf{G}} = \hat{\mathbf{G}}_0\left(\hat{\mathbf{I}} - \hat{\mathbf{t}}\,\hat{\mathbf{G}}_0\right). \qquad (2.2.43)$$

Due to their cornerstone value for the calculational procedure equations (2.2.42) and (2.2.43) are often called *fundamental equations of the multiple scattering theory*. The electronic eigenvalue spectrum of the problem is then completely defined by the singularities of the Green's function of the perturbed system:

$$\det\left[\hat{\mathbf{t}}^{-1} - \hat{\mathbf{G}}_0\right] = 0 \qquad (2.2.44)$$

2.2.8 Spherical representation of the scattering problem

Let us now proceed towards the description of realistic systems. Crystalline solids can be very well characterized as an arrangement of atom-centered perturbing potentials

$$V_{eff}(\mathbf{r}) = \sum_n V_n(\mathbf{R}^n + \mathbf{r}'), \qquad (2.2.45)$$

where $\mathbf{R}^n$ are the coordinates of the n-th atom. Let us consider the perturbation of a free space (or an environment with a constant potential) with a single spherical atomic potential V_s. For the reference free-electron system the Hamiltonian contains only the kinetic energy term and the eigenfunctions of the system are plain waves. The corresponding Green's function has the form: [137, 138]

$$g(\mathbf{r}, \mathbf{r}', \epsilon) = -\frac{1}{4\pi} \frac{e^{ik|\mathbf{r}-\mathbf{r}'|}}{|\mathbf{r}-\mathbf{r}'|}, \qquad (2.2.46)$$

where $k = \sqrt{\epsilon}$. In regard of the fact that we are dealing with a central potential, it is much more convenient to work in angular-momentum representation. The Green's function can then be rewritten as:

$$g(\mathbf{r}, \mathbf{r}', \epsilon) = \sum_L Y_L(\mathbf{r}) g_l(r, r'; \epsilon) Y_L(\mathbf{r}') \qquad (2.2.47)$$

$$g_l(r, r'; \epsilon) = -i\sqrt{\epsilon} j_l(\sqrt{\epsilon} r_<) h_l(\sqrt{\epsilon} r_>) \qquad (2.2.48)$$

where $L \equiv (l, m)$ is the combined index, $Y_L = Y_{lm}$ are the real spherical harmonics, j_l, n_l and $h_l = j_l + i n_l$ are spherical Bessel, Neumann and Hankel functions respectively and

$$r_< \equiv \min\{r, r'\} \quad , \quad r_> \equiv \max\{x, x'\}.$$

Let us assume, that the scattering potential spherically symmetric and of a finite range R_S [129, 133, 134, 135]:

$$V_s(r) = \begin{cases} V(r), & r \leqslant R_S \\ 0, & r > R_S \end{cases} \qquad (2.2.49)$$

Then the radial wavefunctions $R_l(r; \epsilon)$ satisfy the solution of the Radial Schrödinger equation

$$\left[-\frac{1}{r} \frac{\partial^2}{\partial r^2} r + \frac{l(l+1)}{r^2} + V_s(r) - \epsilon \right] R_l(r; \epsilon) = 0. \qquad (2.2.50)$$

The asymptotic form for $R_l(r; \epsilon)$ at $r \to \infty$ is

$$R_l(r; \epsilon) \to \frac{A_l}{\sqrt{\epsilon} r} \sin\left(\sqrt{\epsilon} r - \frac{l\pi}{2} + \delta_l(\epsilon)\right), \qquad (2.2.51)$$

where A_l is a constant and $\delta_l(\epsilon)$ is the phase shift of the solution with respect to the wavefunction for a vanishing potential.

For $r > R_S$, where $V_s(r) = 0$, the genegal solution of the radial equation is a superposition of two linearly independent radial solutions:

$$R_l(r; \epsilon) = B_l j_l(\sqrt{\epsilon} r) + C_l n_l(\sqrt{\epsilon} r), \qquad (2.2.52)$$

2.2 Green's function method

where B_l and C_l are constants. Using the asymptotic form of Bessel functions we obtain from (2.2.51), (2.2.52) and the Lippmann-Schwinger equation the following representation for $R_l(r;\epsilon)$: [139]

$$R_l(r;\epsilon) = j_l\left(\sqrt{\epsilon}r\right) - i\sqrt{\epsilon}\,t_l(\epsilon)\,h_l\left(\sqrt{\epsilon}r\right), \quad r > R_S \tag{2.2.53}$$

where t_l is the matrix element representation of the t-operator: [139]

$$t_l(\epsilon) = \int_0^{R_S} j_l\left(\sqrt{\epsilon}r'\right) V_s(r') R_l(r';\epsilon) r'^2 dr', \tag{2.2.54}$$

or related to the phase shift δ_l:

$$t_l(\epsilon) = -\frac{1}{\sqrt{\epsilon}} \sin\delta_l(\epsilon) e^{i\delta_l(\epsilon)} \tag{2.2.55}$$

This solution is of a regular character, i.e. converging at $r \to 0$. The irregular solution, i.e. diverging at $r \to 0$ will have the form [139]

$$H_l(r;\epsilon) = h_l\left(\sqrt{\epsilon}r\right), \quad r > R_S \tag{2.2.56}$$

The incoming and scattered waves can now be expanded into Bessel functions and spherical harmonics as:

$$\psi_k^{inc}(\mathbf{r}) = \sum_L a_{kL}^0 j_l(\sqrt{\epsilon}r) Y_L(\mathbf{r}), \quad r > R_S \tag{2.2.57}$$

$$\psi_k^{sc}(\mathbf{r}) = \sum_L a_{kL}^{sc} j_l(\sqrt{\epsilon}r) Y_L(\mathbf{r}), \quad r > R_S \tag{2.2.58}$$

Inside the sphere, where the potential is non-vanishing, we have

$$\psi_k(\mathbf{r}) = \sum_L a_{kL} R_L(r,\epsilon) Y_L(\mathbf{r}), \quad r < R_S. \tag{2.2.59}$$

With the boundary condition for $R_L(r,\epsilon)$ given at $r = R_S$ by the continuity of the wave function, we obtain

$$a_{kL}^{sc} = -i\sqrt{\epsilon}\,t_l(\epsilon)\,a_{kL}^0. \tag{2.2.60}$$

Finally, the Green's function for the scattering problem of a single central potential, can also be written as a product of two linearly independent solutions $R_l(r;\epsilon)$ and $H_l(r;\epsilon)$ of the radial equation [139, 129, 133, 134, 135, 136]:

$$\begin{aligned} G_s(\mathbf{r},\mathbf{r}';\epsilon) &= -i\sqrt{\epsilon} \sum_L R_l(r_<;\epsilon) H_l(r_>;\epsilon) Y_L(\mathbf{r}) Y_L(\mathbf{r}') \\ &= \sum_L G_l(r,r',\epsilon) Y_L(\mathbf{r}) Y_L(\mathbf{r}'). \end{aligned} \tag{2.2.61}$$

The regular and irregular solutions must satisfy the Wronski relation:

$$[H_l(r;\epsilon), R_l(r;\epsilon)] = \frac{1}{r^2\sqrt{\epsilon}} \tag{2.2.62}$$

In practice, equation (2.2.50) is usually integrated numerically from $r = 0$ to $r = R_S$ to obtain R_l and the requirement of the continuity of the logarithmic derivative of the solution yields the t-matrix elements. Analogously, the integration from $r = R_S$ inwards gives one the diverging radial wave function H_l given the boundary condition (2.2.56) at $r = R_S$. [139, 134, 135]

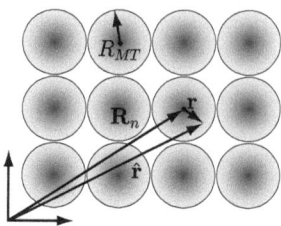

Figure 2.1: A crystal reference and a cell-centered coordinate systems.

2.2.9 Multiple scattering: bulk crystal

As already mentioned, bulk solids can be very well described by an effective potential comprised of an arrangement of identical single-atom potentials. Let us divide the volume of the solid into disjoint space domains, as defined in (2.2.32), so that each domain contains one charge-neutral atom. For example, such division might be achieved by a Wigner-Seitz (WS) construction. Then each WS cell might be associated with an atomic potential defined within its boundaries. A system might be further simplified if we assume that each potential is nonvanishing only in a sphere of radius R_S centered at the atom. There are several approaches to selecting the radius of the sphere. The simplest approximation is to assume atomic spheres to have the same radii R_{ASA}. This approach is called the *atomic spheres approximation* (ASA).

Here we will adopt a slightly more general approach, named the *muffin-tin* (MT) approximation. It is assumed that the scattering potential is spherically symmetric around each scattering center (atomic site) within a sphere of radius R_{MT} and constant otherwise (without limiting the generality, the interstitial potential can be taken to be zero for convenience) [see (2.2.49)]. The spheres are constructed to touch each other but are assumed to be non-overlapping.

We will further use the approximation of an ideal crystal, which means that for any lattice vector $\mathbf{R}_i$ the following relation for the effective potential applies:

$$V_{eff}(\mathbf{r} + \mathbf{R}_i) = V_{eff}(\mathbf{r}). \quad (2.2.63)$$

For further calculational convenience let us assume the cell-centered coordinate system (Fig. 2.1):

$$\hat{\mathbf{r}} \to \mathbf{R}^n + \mathbf{r}$$
$$\hat{\mathbf{r}}' \to \mathbf{R}^{n'} + \mathbf{r}' \quad (2.2.64)$$
$$|\mathbf{r}|, |\mathbf{r}'| \leqslant R_{MT}.$$

The Green's function is defined by the following Kohn-Sham equation

$$\left[-\nabla^2 + V_n(r) - \epsilon\right] G(\mathbf{r} + \mathbf{R}^n, \mathbf{r}' + \mathbf{R}^{n'}; \epsilon) = -\delta_{nn'}\delta(\mathbf{r} - \mathbf{r}'). \quad (2.2.65)$$

For $n \neq n'$ the Green's function satisfies the homogeneous Schrödinger equation and can be expanded in it's independent regular solutions $R_L^n(\mathbf{r}; \epsilon)$ and $R_L^{n'}(\mathbf{r}; \epsilon)$. In the muffin-tin approximation they can easily be expressed as the solutions of Eq. (2.2.50) $R_L^n(\mathbf{r}; \epsilon) = R_l^n(\mathbf{r}; \epsilon) Y_L(\mathbf{r})$. For $n = n'$ we have a Green's function for a central potential problem as shown

2.2 Green's function method

in Eq. (2.2.61) in addition with a boundary condition of back-scattering by all the other potentials in the crystal. Thus the total solution can be expressed as a sum of a general solution of a homogeneous equation (2.2.65) and a special solution of its inhomogeneous version. In the mixed site-angular-momentum representation it would look like:

$$G(\mathbf{r}+\mathbf{R}^n, \mathbf{r}'+\mathbf{R}^{n'}; \epsilon) = -i\sqrt{\epsilon}\sum_L R_L^n(\mathbf{r}_<; \epsilon) H_L^n(\mathbf{r}_>; \epsilon) \delta_{nn'} + \\ + \sum_{LL'} R_L^n(\mathbf{r}; \epsilon) G_{LL'}^{nn'}(\epsilon) R_{l'}^{n'}(\mathbf{r}; \epsilon), \qquad (2.2.66)$$

where $H_L^n(\mathbf{r}; \epsilon) = H_l^n(r; \epsilon) Y_L(\mathbf{r})$ it the irregular solution of the radial Schrödinger equation at the atomic cell n. The coefficients $G_{LL'}^{nn'}$ are usually called the *structural Green's functions*. To obtain them we have to turn once again to the Dyson equation, taking free space as the unperturbed system:

$$G(\mathbf{r}+\mathbf{R}^n, \mathbf{r}'+\mathbf{R}^{n'}; \epsilon) = g(\mathbf{r}+\mathbf{R}^n, \mathbf{r}'+\mathbf{R}^{n'}; \epsilon) + \\ + \sum_{n''} \int g(\mathbf{r}+\mathbf{R}^n, \mathbf{r}''+\mathbf{R}^{n''}; \epsilon) V_{n''} G(\mathbf{r}''+\mathbf{R}^{n''}, \mathbf{r}'+\mathbf{R}^{n'}; \epsilon) d\mathbf{r} . \qquad (2.2.67)$$

after substituting (2.2.66) into (2.2.67), expanding the free-space Green's function and some mathematical manipulations [140, 139, 134], we obtain the *algebraic Dyson equation* determining the structural Green's functions:

$$G_{LL'}^{nn'}(\epsilon) = g_{LL'}^{nn'}(\epsilon) + \sum_{n'',L''} g_{LL''}^{nn''}(\epsilon) t_{l''}^{n''}(\epsilon) g_{L''L'}^{n''n'}(\epsilon) \qquad (2.2.68)$$

in which the t-matrix replaces the normally appearing in a Dyson equation potential.

In practice it is convenient to utilize the translational invariance on an ideal crystal lattice and calculate the structural Green's functions in the **k**-space. The transition can be made by the following Fourier transformation:

$$G_{LL'}(\mathbf{k}; \epsilon) = \sum_{n'} G_{LL'}^{nn'}(\epsilon) e^{-i\mathbf{k}\cdot(\mathbf{R}^n - \mathbf{R}^{n'})}. \qquad (2.2.69)$$

The algebraic Dyson equation (2.2.68) then takes the form:

$$G_{LL'}(\mathbf{k}; \epsilon) = g_{LL'}(\mathbf{k}; \epsilon) + \sum_{L''} g_{LL''}(\mathbf{k}; \epsilon) t_{l''}(\mathbf{k}; \epsilon) g_{L''L'}(\mathbf{k}; \epsilon), \qquad (2.2.70)$$

where the t-matrix also looses its n-dependence due to the translational invariance. In real calculations the structural Green's functions are considered as matrices in (L, L') space and solved by matrix inversion. The number of considered l-s is usually limited by a cutoff $l = l_{max}$ for which the t-matrix becomes negligibly small. For calculation of transitional-metal systems $l_{max} = 3$ or 4 is usually sufficient. After the inverse fourier transform the structural coefficients $G_{LL'}^{nn'}(\epsilon)$ in the KKR matrix notation take the shape:

$$G_{LL'}^{nn'}(\epsilon) = \frac{1}{V_{BZ}} \int_{BZ} e^{-i\mathbf{k}\cdot(\mathbf{R}^n - \mathbf{R}^{n'})} \left[\left(1 - \hat{\mathbf{g}}(\mathbf{k}; \epsilon)\hat{\mathbf{t}}(\epsilon)\right)^{-1} \hat{\mathbf{g}}(\mathbf{k}; \epsilon)\right]_{LL'} d^3\mathbf{k} , \qquad (2.2.71)$$

where the integration is carried out over the Brillouin zone volume V_{BZ}. The local values at site n, like the charge density or the local density of states, can be obtained from a single term for $n = n'$: $G_{LL'}^{nn}$.

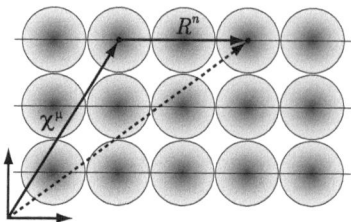

Figure 2.2: Coordinate notations for the description of the Green's function of layered systems.

2.2.10 Surfaces and layered systems

The above presented approach can be easily extended to treat layered or semi-infinite systems [140, 139, 134]. The main difference is that along one direction the translational invariance is broken leaving only a two-dimensional periodicity, parallel to a surface or an interface. The Fourier transformations are now held within a two-dimensional surface Brillouin zone (SBZ), and the integration is carried out over all $\mathbf{k}_\parallel$ wave vectors in the SBZ [140, 139, 134]:

$$G^{n\mu,n'\mu'}_{LL'}(\epsilon) = \frac{1}{A_{SBZ}} \int_{SBZ} e^{i\mathbf{k}_\parallel \cdot (\mathbf{R}^n - \mathbf{R}^{n'})} e^{i\mathbf{k}_\parallel \cdot (\chi^\mu - \chi^{\mu'})} \\ \times \left[\left(1 - \hat{\mathbf{G}}^r(\mathbf{k}_\parallel;\epsilon) \Delta \hat{\mathbf{t}}(\epsilon)\right)^{-1} \hat{\mathbf{G}}^r(\mathbf{k}_\parallel;\epsilon) \right]^{\mu\mu'}_{LL'} d^2\mathbf{k}_\parallel , \quad (2.2.72)$$

where $\mathbf{R}^n$ are the in-plane cite vectors in the two-dimensional lattice and χ^μ point to atomic coordinates in different layers (see Fig. 2.2). A_{SBZ} is the area of the surface Brillouin zone.

For surface calculations ideal crystal is considered to be perturbed by vacuum. The structure of the lattice remains unchanged, just the corresponding potentials are replaced by vacuum ones (empty spheres). To obtain realistic surface properties one can either choose to perturb an infinite crystal by a slab of a finite size, or to terminate the crystal completely and apply half-infinite boundary condition. In the first approach it is usually sufficient to take into account twice 3 or 4 vacuum layers to obtain realistic density of states at the surface. Such fast decay of the electronic properties' coherence also allows one to considerably cut the calculational effort by applying the so-called *screening procedure*.

2.2.11 Screened KKR

Screened of *tight-binding* KKR formalism allows one to calculate large systems as it assures that the computational effort for matrix inversion, while solving the algebraic Dyson equation, ideally grows linearly $O(N)$ with the size of the system (number of inequivalent atoms N in the unit-cell). In conventional KKR approach the same results can be obtained only with an effort of $O(N^3)$ [140, 139, 134].

Such computational profit is made possible by an optimal choice of a new reference system, which assures that the Green's function falls of exponentially with increasing distance from the scattering center. This leads to an effective decoupling of distant atomic sites, which gave the transformation its *screening* name. The occasional name "tight-binding KKR" originates from the resemblance to the tight-binding approach in treating atomic orbitals.

2.2 Green's function method

Previously we have always chosen as a reference for the Dyson equation the free space. It can, however, be shown [140, 139] that a choice of a different reference system for the same "real" lattice structure leads to the same form of the Dyson equation with a difference of t-matrices between the real and reference systems $(t_{LL'}^n(\epsilon) - t_{LL'}^{r,n}(\epsilon))$ taking the place of the $t_{LL'}^n(\epsilon)$ in equation (2.2.68):

$$\hat{\mathbf{G}}(\epsilon) = \hat{\mathbf{G}}^r(\epsilon) + \hat{\mathbf{G}}^r(\epsilon)\mathbf{\Delta}\hat{\mathbf{t}}(\epsilon)\hat{\mathbf{G}}(\epsilon) \,, \qquad (2.2.73)$$

where $\mathbf{\Delta}\hat{\mathbf{t}}(\epsilon) = \hat{\mathbf{t}}(\epsilon) - \hat{\mathbf{t}}^r(\epsilon)$.

The main problem of the free space as a reference system is that it brings with it additional singularities, corresponding to free-electron eigenstates in the interstitial potential-free regoins. Thus, our task is to devise a reference system with no eigenstates in the energy range of free electrons. A choice of such a system as a reference would secure the exponential decay of the Green's function [141, 142].

In principle, any potential without eigenstates in the energy range of free electrons can be a candidate for the new reference system, but the simplest choice, probably, is an infinite periodic arrangement of finite-height repulsive potentials V_n^R constant within the muffin-tin spheres of radius R_{MT}^n around each scattering center R^n:

$$V_s^R(\mathbf{r}) = \begin{cases} V_C & , r \leqslant R_{MT}^n \\ 0 & , \text{otherwise} \end{cases} , \qquad (2.2.74)$$

where V_C is a constant, usually chosen to be a few Rydberg. It has been predicted and computationally proved that for such a potential the eigenstate spectrum starts above an energy of E_{bot} which is slightly bellow the barrier height V_C. For example, for a barrier of 1 Ry and 2 Ry the eigenstates spectrum bottom lie at 0.7 and 1.35 Ry respectively [141, 142].

The above described procedure can be effectively applied to great number of various systems. SKKR has proved to work with insulators, semiconductors and ,in particular, with metals.

2.2.12 Impurities

The domain of KKR's effectiveness does not end with layered systems and interfaces. It can equally well be applied to single impurities in the bulk and at surfaces. Such structures essentially break the translational symmetry and thus have to be treated within the real-space representation of the KKR formalism. For instance, the structural Green's function of an ideal surface can be cast into the real space representation and then used as a reference for the calculation of a surface-impurity problem. A corresponding algebraic Dyson equation would then look like:

$$\hat{\mathbf{G}}_{LL'}^{nn'}(\epsilon) = \hat{\mathbf{G}}_{LL'}^{0,nn'}(\epsilon) + \sum_{n'',L''} \hat{\mathbf{G}}_{LL''}^{0,nn''}(\epsilon) \, \mathbf{\Delta t}_{l''}^{n''}(\epsilon) \, \hat{\mathbf{G}}_{L''L'}^{n''n'}(\epsilon) \,, \qquad (2.2.75)$$

where $\hat{\mathbf{G}}_{LL'}^{nn'}(\epsilon)$ and $\hat{\mathbf{G}}_{LL'}^{0,nn'}(\epsilon)$ are the the energy-dependent structural Green's functions for the surface-impurity problem and for a clean surface correspondingly, and t-matrix $\mathbf{\Delta t}_l^n(\epsilon)$ accounts for the difference in the scattering properties of site n induced by the presence of an impurity potential.

For the first time self-consistent calculations of single Ni, Zn, Ga and Ge impurities in Cu crystal were performed in 1979 by Zeller and Dederichs [143]. In a year, realistic self-consistent electronic structures of 3d magnetic impurities embedded into bulk Cu and Ag were presented [144]. Obtained results were in a qualitative agreement with the Anderson model, but it was emphasized that modifications of the impurities' electronic structures due to host band structure were important. Local magnetic moments were in a reasonable agreement with available experimental data. Further improvements of the method made possible calculations of the exchange interaction of magnetic dimers in nonmagnetic hosts like Cu or Ag [108].

Development of the KKR for layered systems open up a possibility to study clusters on surfaces. Calculations performed for 4d transition-metal clusters on Ag(001) substrate, contrary to nonmagnetic bulks of the species investigated, revealed strong tendency to magnetism [145, 146]. Exact magnetic moments of 4d nanostructures were found to depend strongly on geometries of clusters. Later on similar calculations were conducted for 3d, 4d and 5d transition metals on Pd(001) and Pt(001) substrates [147].

2.3 Force theorem

Despite the fact that the ever increasing processing power of modern computers allowed to facilitate the calculation of large atomic systems, some computational aspects still remain a problem. For example the digital precision used in calculations has remained largely unchanged in the last several decades. This means, that even having the power to perform fully selfconsistent calculations we may loose some physical effects to computational precision errors. As an example, the calculation of the indirect interaction between single impurities in a metallic or semiconductor host or on a surface can be given. The total energy of a host system with two interacting impurities E_{1+2} can be expressed as

$$E_{1+2} = E_{host} + E_1 + E_2 + E_{int} , \qquad (2.3.1)$$

where E_{host} is the energy of a clean host (surface or bulk), E_1 and E_2 are the total energies of the two adsorbates and E_{int} encapsulates the interaction energy between the two impurities. Thus the interaction energy alone can be expressed as

$$E_{int} = E_{1+2} - E_{host} - E_1 - E_2 . \qquad (2.3.2)$$

If total energies derived from the KKR method are used in this formula, then the resulting accuracy of interaction energy calculations for metallic systems usually falls below $10^{-2} - 10^{-3}$ eV which effectively leaves the indirect interaction energy of two impurities beyond the calculational precision.

To circumvent this problem the so-called *force theorem* can be applied. It says that the total energy difference can also be calculated (within the mistake margin of the second order of magnitude) from the single-particle energies alone, which yields a precision of several orders of magnitude higher that in the total energy approach [108]

The arguments are based on a *frozen potential approximation* which was initially proposed in a different context by Pettifor (1977), Varma and Pettifor (1979) and others (Methfessel and Kübler (1982), Skiver (1982)).

We can express the total energy of the system as the sum of two contributions: the single particle term E_{sp} and the double counting contribution E_{dc}

$$E_{tot} = E_{sp} + E_{dc} , \qquad (2.3.3)$$

2.3 Force theorem

where (for a general case of a magnetic impurity)

$$E_{sp} = \int_{-\infty}^{\epsilon_F} d\epsilon (\epsilon - \epsilon_F)(n^+(\epsilon) + n^-(\epsilon)) \qquad (2.3.4)$$

$$E_{dc} = -\int d\mathbf{r} \left(n^+(\mathbf{r}) V_{eff}^+(\mathbf{r}) + n^-(\mathbf{r}) V_{eff}^-(\mathbf{r}) \right) + W\{n^+(\mathbf{r}), n^-(\mathbf{r})\}. \qquad (2.3.5)$$

Here W represents the sum of the Coulomb and the exchange correlation energies, $V_{eff}^{\pm}(\mathbf{r})$ are the perturbing potentials from which the spin densities $n^{\pm}(\mathbf{r})$ and densities of states $n^{\pm}(\epsilon)$ are derived. Due to the extremal properties, the $E\{n^+, n^-\}$ error contribution is of the second order [108].

Another useful extremal property is that of the double counting energy E_{dc}. For fixed perturbing potentials the double counting energy $E_{dc}\{n^+, n^-\}$ is insensitive to variations in $n^{\pm}(\mathbf{r})$ [108]. In the first order it can be written

$$\delta E_{dc}\big|_{V_{eff}^{\pm}} = \int d\mathbf{r} \left[\left(\frac{\delta W}{\delta n^+(\mathbf{r})} - V_{eff}^+(\mathbf{r}) \right) \delta n^+(\mathbf{r}) + \left(\frac{\delta W}{\delta n^-(\mathbf{r})} - V_{eff}^-(\mathbf{r}) \right) \delta n^-(\mathbf{r}) \right]. \qquad (2.3.6)$$

Since the exact solution requires $V_{eff}^{\pm} = \delta W \delta n^p m(\mathbf{r})$, the error in equation (2.3.6) is indeed of the second order. Therefor the double counting contribution can be calculated from approximate spin densities $\tilde{n}^{\pm}(\mathbf{r})$.

We can take advantage of the above mentioned extremal properties and choose $V_{eff}^{\pm}(\mathbf{r})$ and $\tilde{n}^{\pm}(\mathbf{r})$ in the following way. We set the potential to be a superposition of the clean host potential V_{eff}^0 and the additional impurity terms ΔV_{eff}^1 and ΔV_{eff}^2:

$$V_{eff} = V_{eff}^0 + \Delta V_{eff}^1 + \Delta V_{eff}^2, \qquad (2.3.7)$$

which is a first order approximation for weakly interacting impurity potentials. Furthermore we can assume that $\Delta V_{eff}^{1,2}$ is strongly localized at impurity sites (although this might not necessarily be the case).

For the spin densities $\tilde{n}^{\pm}(\mathbf{r})$ it is convenient, as well as reasonable, to assume that their sum, the total charge density $\tilde{n} = \tilde{n}^+ + \tilde{n}^-$, is the same for both spin configurations. Since the charge density depends only quadratically on the magnetization and since we assume both impurities to have the same absolute magnetic moments (for both configurations), this again is an adequate first-order estimate. This approximation allows one to avoid the necessity to deal with inequivalent Coulomb integrals of the double counting energies as they are (in this approximation) identical and cancel each other out when the magnetic interaction is evaluated. The remaining, genuinely magnetic, part E_{dc}^m of the double counting energy depends at least quadratically on the magnetization. In the local density approximation it can be written in the form [108]:

$$E_{dc}^m = \int d\mathbf{r} \left(f_2(n(\mathbf{r})) m^2(\mathbf{r}) + f_4(n(\mathbf{r})) m^4(\mathbf{r}) + ... \right). \qquad (2.3.8)$$

Since the magnetization is well localized, in most cases one obtains important contributions only from the two impurity cells. Our *ab initio* calculations along with numerous other publications (see, e.g., [148]), show that in transitional metal hosts, for instance Cu or Ag, the moment on a neighboring sites is less than 1% of the local impurity moment, so that the contributions from all neighboring sites are about 10^{-3} times smaller than the contributions to (2.3.8) from the impurity site [108]. If, moreover, one replaces the magnetization on each site by the magnetization $m^o(\mathbf{r})$ innate to a single impurity, one would also see that the magnetic part E_{dc}^m of the double counting energy is the same for each configuration. Linear terms proportional to $\Delta m = m - m^0$ do not occur due to the extremal property of E_{dc}^m (equation (2.3.6)) [108].

Thus one can conclude that in the first order the magnetic interaction energy is given by the difference of the single-particle energies for both configurations, provided a frozen single impurity potential is chosen. The approximation is the better for a weaker interaction and becomes exact asymptotically. It is, however, important to understand, that this proof applies only to the difference in interaction energy between the two configurations, but not to the interaction energies themselves [108].

2.4 KKR GF Calculations: workflow and examples

To present the actual calculational workflow and illustrate the application of the KKR Green's function method to the calculation of electronic properties of surface structures we present in this section an example calculation for a simple system: a single Co adatom on a Cu(111) surface.

2.4.1 Bulk of an *fcc* copper crystal

Our KKR calculations start with an infinite Cu *fcc* crystal, which can be easily obtained as a 3D periodic perturbation of a free space or vacuum[1]. To match experimental conditions we use the experimental lattice constant of 3.615 Å. In the iterative procedure (see Fig. 2.3a) the Dyson equation (2.2.12) is solved for a set of k-points covering the irreducible part of 3D Brillouin Zone (BZ). To obtain such a set, various strategies can be followed. In our calculations we use one of the most widely used: the Monkhorst-Pack method [150], which allows one to generate a sets of special points in the Brillouin zone which provides an efficient means of integrating periodic functions of the wave vector. To achieve reasonable precision in transitional metal calculations about $10^3 - 10^6$ k-points in the irreducible part of 3D BZ are needed. Upon solving the Dyson equation one obtains the Green's function of our initial system, perturbed by a lattice of atomic potentials, from which energy and momentum space resolved electronic densities can be extracted. The ground state electronic density can be obtained in a self-consistent cycle, when the resulting electronic density is used as an input for the next solution run of the Dyson equation. If densities calculated at several consequent steps of the self-consistent cycle are the same within a predefined errorbar (see Fig. 2.3a), one can assume to have obtained a good estimation of the ground state electronic density. The example of the calculated total density of states of a Cu bulk is presented in Fig. 2.3b. It is in ample agreement with the densities calculated by other methods (see, f.e. Fig. 2.3c where the calculated LDOS from [149] is presented). The Fermi level is defined in a usual way, by counting the valence electron states up to the chemical potential in the unit cell.

2.4 KKR GF Calculations: workflow and examples

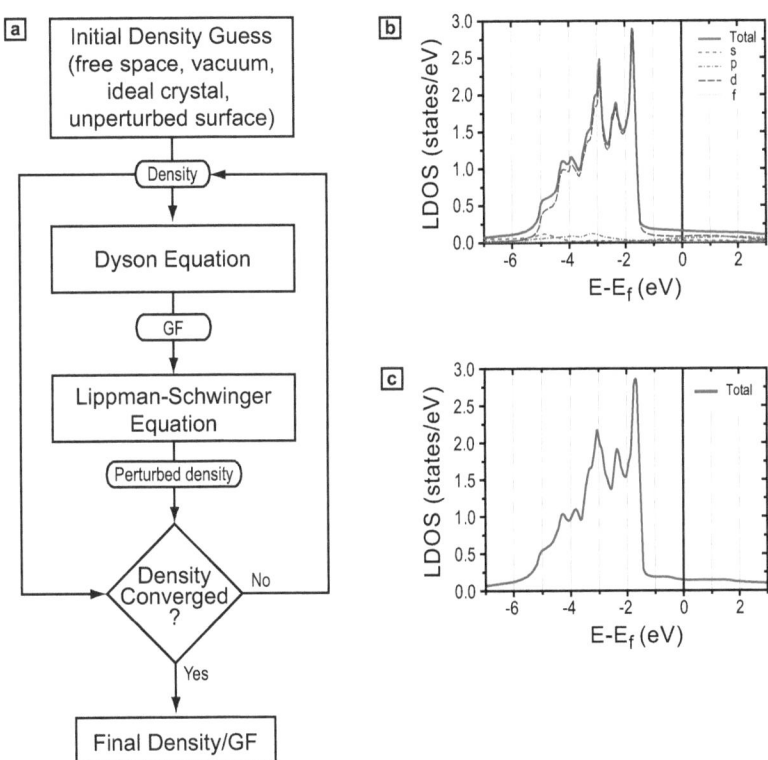

Figure 2.3: (a) Recurrent algorithm for finding the ground state density and Green's function in the KKR GF approach. (b) Total local density of states of a copper bulk (thick solid red line) and its decomposition into spherical harmonics: s (thin short-dashes red), p (dash-dotted green), d (lond-dashed dark-blue) and f (dotted cyan). (c) Cu LDOS from [149] for comparison.

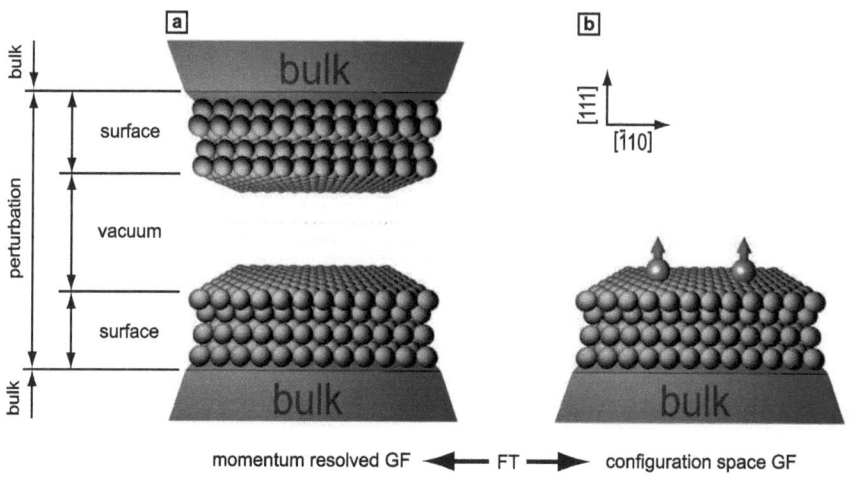

Figure 2.4: (a) Surface represented as a 2D perturbation of the bulk crystal. (b) Impurity is treated as a perturbation of a clean surface. For that the Green's function if Fourier transformed from the momentum space into the configuration space representation.

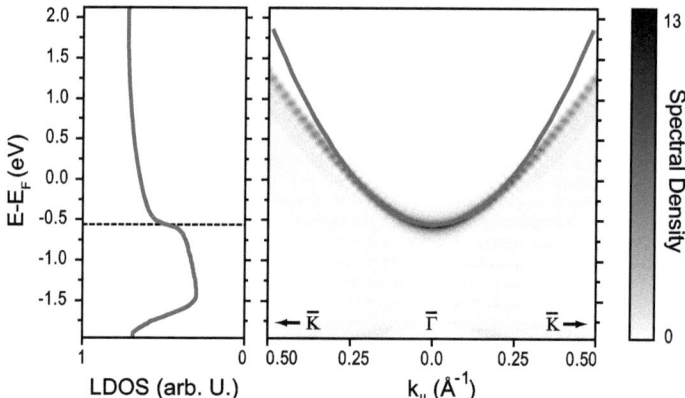

Figure 2.5: Spectral density map (right panel) of the electrons in the first vacuum layer (some 2 Å above the surface). The red curve represents a parabolic fit of the surface state band. The left panel represents the k-integrated local density of states, which shows a characteristic onset in the region of the surface state band bottom. The dashed line in the left panel marks the energy at which the surface state electron density is compared for different surface layers in Fig. 2.6.

2.4 KKR GF Calculations: workflow and examples

Table 2.1: Surface state band parameters obtained in our calculations and in ARPES experiments by Reinert *et al.* [152]. To obtain the parameters the band is fitted to a parabolic model $E(k_\parallel) = E_0 + \hbar^2 \cdot k_\parallel^2/2m^*$, where E_0 is the band bottom position below the Fermi level, m^* is the effective mass of a surface state electron and $\mathbf{k}_\parallel$ is the in-plane wave vector.

	E_0 (meV)	m^*/m_e	k_F (Å^{-1})
KKR	551	0.380	0.231
ARPES	435	0.412	0.215

2.4.2 Cu(111) surface

To obtain a surface we follow the recipe given in Subsection 2.2.10. We perturb the ideal crystal in 2D with a slab of vacuum. Figure 2.4a gives a rough sketch of the resulting geometry. We use at least 6 layers of vacuum to efficiently decouple the two resulting semiinfinite crystals. To guarantee, that the electronic properties of the surface are accurately reproduced and the charge redistribution is properly taken into account we allow at least 4 metal layers at each interface to relax electronically (or, in some cases, also geometrically). Bulk electronic states are introduced in the system as boundary conditions through the scattering matrix $\Delta \hat{t}(\epsilon)$ obtained in the bulk calculations. In accordance with Subsection 2.2.10 we are now formulating our Green's function in the in-plane momentum space. The Green's function $\hat{\mathbf{G}}^r(\mathbf{k}_\parallel; \epsilon)$ now depends on the in-plane momentum $\mathbf{k}_\parallel$ and energy ϵ, which makes is extremely easy for us to obtain the space and energy resolved density of states by just taking the Green function's imaginary part $-1/\pi \, \mathfrak{Im} \hat{\mathbf{G}}^r(\mathbf{k}_\parallel; \epsilon)$. This quantity is usually called the spectral density. It gives one insight into the band structure of the studied system. For instance, the spectral density in the vacuum space above the surface surface would yields valuable information about the nature of surface electrons. A cut of the such a spectral density, calculates some 2 Å above the copper-vacuum interface, along the $\overline{K} - \overline{\Gamma} - \overline{K}$ line of the 2D BZ in the energy range of $[-2.0 \text{ eV}; 2.0 \text{ eV}]$ is presented in Fig. 2.5. One can clearly observe a single dispersive band centered at the $\overline{\Gamma}$-point. This band is a typical signature of a surface state (see Subsection 1.7.5). The nearly parabolic dispersion (the red line in Fig. 2.5 represents a parabolic fit of the band) is a consequence of the quasi-free nature of the electronic localization in the surface plane. Surface state parameters extracted from the fit are presented in Table. 2.1 along with typical values obtained in angle resolved photoemission spectroscopy (ARPES) experiments [151, 152]. It might be noticed, that at larger k-values the actual dispersion curve actually deviates from the ideal parabolic behavior. As was shown in [151] it can be fully attributed to the fact that the surface potential is not flat for the surface state electrons, but represents a periodic lattice of palpable perturbations, which cause a gap to appear at the edges of the BZ. The fact, that KKR yields a very accurate description of the surface electron behavior, will be extremely important in Section 3.3 for the evaluation of the surface-state mediated RKKY-like magnetic interaction of single adatoms on nanoscopic islands.

The surface state electrons are bound to the surface, i.e. their density decays exponentially both into the vacuum and into the crystal. To be sure that KKR is equally precise in the description of this behavior we present the spectral density of electrons at the $\overline{\Gamma}$-point at

---
[1] The Green's function of a free space can be formulated analytically.

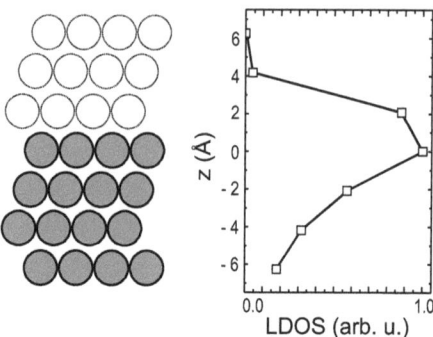

Figure 2.6: The spectral density of electrons at the $\overline{\Gamma}$-point at the surface state band bottom energy (marked by a dashed line in the left panel of Fig. 2.5) for different layers of the Cu(111) surface. The figure has been adopted from [135].

the surface state band bottom energy (marked by a dashed line in the left panel of Fig. 2.5) for different layers of our surface in Fig. 2.6. The presented spectral density is obviously maximal at the interface and rapidly decays into the vacuum as well as into the surface. This gives us additional confidence in the capability of the KKR GF method to describe surface-state-related phenomena.

2.4.3 A single Co adatom on a Cu(111) surface

To calculate an arbitrary nanostructure on a surface in the framework of KKR one has to consider, that the dyson equation can not be formulated for such a structure in the momentum space notation because of the symmetry break induced by the new geometry of the system. It is thus necessary for further calculations to fourier-transform the Green's function of a clean surface into the configuration-space notation. Contrary to the momentum space, the complexity of a real-space formulated Green's function grows with increasing perturbation volume of the system. Consequently, it is paramount to limit the considered system to a necessary minimum. On a metallic surface the potential of an impurity is strongly screened by the conduction electrons of the host. For KKR this means, that the perturbation of the potential is felt at the impurity site and in several empty sites around it. In fact, for most systems, including just the nearest neighbor sites (atomic spheres or Wigner-Seitz cells) is sufficient to produce an accurate result. An example of perturbation geometry, used to calculate a single adatom on a Cu(111) surface, is shown in Fig. 2.7.

After completing the self-consistent calculation cycle (Fig. 2.3) one obtains a configuration-space-formulated Green's function of the system from which then various system's observables can be extracted. Here we present the local density of states at and above the adatom as an example. Figure 2.8a shows the LDOS at a single Co adatom adsorbed on a Cu(111) surface: majority (red, top panel) and minority (blue, bottom panel). One can nicely see the characteristic splitting between of the majority and minority electronic levels. The resulting magnetic moment (1.86 μ_b) is in a good agreement with experimental values and other DFT calculation-methods. The LDOS in vacuum above the Co adatom is presented in Fig. 2.8b:

2.4 KKR GF Calculations: workflow and examples

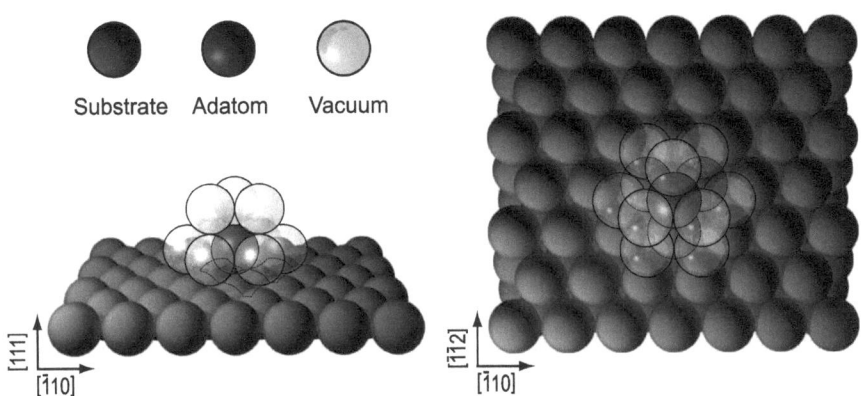

Figure 2.7: Side (left) and top (tight) view of the sample geometry used to calculate a single adatom on a Cu(111) surface. Brown (gray) spheres are the Cu atoms, blue-gray (dark-gray) sphere denotes the impurity and the semitransparent ones denote the vacuum sites. For the actual calculation only solid- or dashed-line-outlined spheres are taken into consideration.

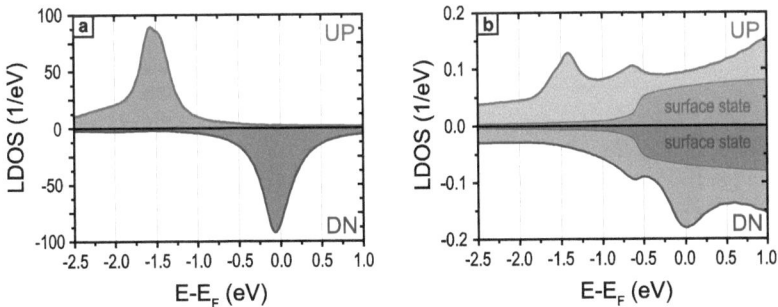

Figure 2.8: (a) LDOS at a single Co adatom adsorbed on a Cu(111) surface: majority (red, top panel) and minority (blue, bottom panel). (b) LDOS in vacuum above a single Co adatom adsorbed on a Cu(111) surface: majority (red, top panel) and minority (blue, bottom panel). Darker shaded areas present the LDOS of a surface state at a clean Cu(111) surface. Results of similar calculation with a higher energy resolution can be found in Chapter 4.

majority (red, top panel) and minority (blue, bottom panel). To relate the picture to a clean Cu(111) surface, the LDOS of a clean surface is presented as darker shaded areas. In detail, various features of the presented local densities of states will be discussed in Chapters 3 and 4.

Summarizing the chapter, we can conclude, that Korringa-Kohn-Rostoker Green's function method is a powerful and efficient tool for describing the electronic and magnetic properties of bulk and surface based metallic systems. It allows for a rapid step-by-step calculation starting from bulk crystals and advancing over interface configurations to bulk- or surface-based impurities. Obtained system parameters are in good agreement with other DFT calculations and experimental values.

Chapter 3
Tailoring the exchange interaction

Despite the tremendous progress in both fundamental and applied surface sciences in the last few decades, the scope of techniques, allowing a precise control over the magnetism of single adatoms is still more than limited (see Sec. 1.1). A considerable part of the present work was dedicated to finding new methods to manipulate the magnetism of single atomic-scale nanostructures adsorbed on metallic surfaces. In this chapter we will discuss several such possibilities.

The technological breakthrough that brought about gigabyte magnetic memory devices was enabled by the discovery of the GMR effect by P. Grünberg and A. Fert [1, 2]. Their fundamental success had triggered a number of consequent studies. Theoretical investigations have revealed that GMR has a common source with another interesting effect, the effect of an oscillatory interlayer exchange coupling (IEC) [1, 10, 11, 12]. Both of them originate in the spin-selective scattering of conductance electrons at magnetic layers. If two magnetic layers are separated with a non-magnetic metallic spacer, the conduction electrons scattered at each magnetic layer are forced to interfere and form standing waves inside the spacer. Since the scattering of minority and majority electrons at magnetic layers is different, the distributions of electronic densities in systems with parallel and antiparallel alignments of the spins should also be different. As a result, the system's ground state energy and spin configuration strongly vary with the thickness of the spacer, resulting in the oscillations of the interlayer exchange coupling [11, 12].

3.1 Coupling of single atomic units to a monolayer across a paramagnetic spacer

In fact, this reasoning is not restricted to the case of monolayers but can be applied to any kind of magnetic impurities coupled via conductance electrons. Affecting the interference of conduction electrons scattered at each impurity, e.g. by introducing structural changes in the metallic host, one can tune the exchange interaction, similarly to the IEC. Nowadays modern experimental techniques such as the scanning tunneling microscopy (STM) make it possible to build in an atom-by-atom fashion complex surface nanostructures with predefined positions of magnetic impurities [153, 154]. In this section we present a novel approach to controlling single spins in nanostructures adsorbed on metallic surfaces. We demonstrate that exchange coupling of adatoms and addimers to a magnetic layer across a nonmagnetic spacer displays an oscillatory behavior. This allows one, by deliberate choice

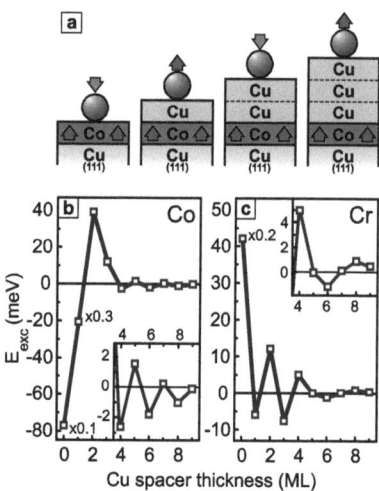

Figure 3.1: (color online) (a) The setup for calculations: an adatom coupled to a Co layer through a nonmagnetic Cu spacer of a varying thickness. Exchange coupling energies of a Co (b) and Cr (c) adatoms versus spacer thickness. First several points of the curves were scaled down for clarity. The scaling factors are given next to respective data points. The insets in each graph show the respective curves on a smaller scale at larger spacer thicknesses.

of the spacer's thickness, to control the magnetic configuration and exchange interaction of single magnetic adatoms, driving them into either a ferro- or an antiferromagnetic behavior or even suppressing their magnetic properties.

To study the interplay between the exchange interaction and the system's geometry we choose a Cu(111) surface as a base for our calculations. Without limiting the generality we consider, as the first and simplest model system, single magnetic 3d adatoms (Co and Cr) placed on top of a Cu spacers of various thicknesses covering a single Co monolayer (ML) (Fig. 3.1a). Our choice of geometry and atomic species is governed by the fact, that thin Co films are known to have out-of-plain magnetization as an inherent property [155]. This fact provides for an increased surface symmetry for the orientation of adatom and addimer spins. The Co layer thickness of 1 monolayer (ML) was chosen as a marginal case and also should not affect the generality of results and conclusions.

The dependence of the exchange interaction energy of an adatom on the thickness of the spacer is shown in Fig. 3.1b for Co and Fig. 3.1c for Cr. The presence of oscillations in the exchange coupling energies, similar to those observed for the interlayer exchange coupling [156, 11], can be attributed to the effect of quantum confinement in the overlayer. The presence of the vacuum barrier on one side of the overlayer and the magnetic layer on the other causes the electrons to become effectively confined between them [121]. Moreover, due to the ferromagnetic nature of the Co monolayer, the confinement of majority and minority electrons will be different, causing the formation of spin-polarized interference patterns. The magnetic properties of adatoms adsorbed on an overlayer are inevitably affected by the changes in the conduction electron densities.

3.1 Coupling of single atomic units to a monolayer across a paramagnetic spacer

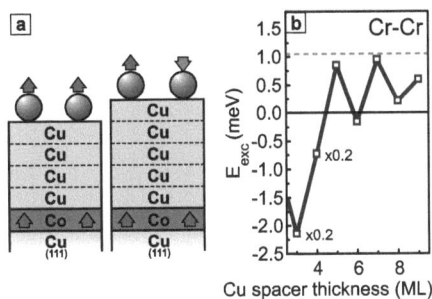

Figure 3.2: (color online). System under consideration: Cr adatoms coupled to a Co monolayer across a Cu spacer of a varying thickness (a). Exchange coupling of a Cr adatoms at 7.66 Å separation aligned along the $[\bar{1}01]$ direction of a Cu(111) surface versus spacer thickness (b). Due to the energetic non-degeneracy of $\uparrow\uparrow$ and $\downarrow\downarrow$ configurations the exchange energies were calculated as follows $E_{exc} = min(E_{\uparrow\uparrow}, E_{\downarrow\downarrow}) - E_{\uparrow\downarrow}$. The gray dashed line gives the level of the exchange coupling strength between Cr atoms on a clean Cu(111) surface. First two points of the curve were scaled down for clarity. The scaling factors are given next to respective data points.

The coupling energies presented in Fig. 3.1(b and c) suggest that by coupling the spins of adatoms to that of a monolayer one gets a reliable means of stabilizing single atomic spins on the surface in either a ferro- or an antiferromagnetic configuration. The switching between configurations can be done by adjusting the thickness of the overlayer.

Let us now move forward, to the second example of spin-dependent scattering affecting the exchange interaction in the system. Up to now we have only considered single adatoms on the surface. If we now add a second adatom to the system, thus creating a dimer (Fig. 3.2a), we will find that the orientation of each of the atomic spins is determined by the competition between two exchange couplings: a coupling to the magnetic monolayer and the interatomic coupling in the dimer. Consequently, precise control over the thickness of the spacer provides us with an additional degree of freedom in adjusting the exchange interaction between single adatoms at any separation. At smaller spacer thicknesses where the coupling of a single adatom to the monolayer prevails the monolayer acts as a stabilizing element, rigidly fixing the dimer in either a $\uparrow\uparrow$ or a $\downarrow\downarrow$ configuration. At larger spacer thicknesses, when the interatomic exchange energy becomes comparable to that of the coupling to the monolayer, the system's spins become most susceptible to manipulations by changing the spacer thickness and interatomic separation. As an example the exchange interaction energy of a Cr-Cr dimer at 7.66 Å separation is shown in Fig. 3.2b as a function of the spacer thickness. It is clear that by adjusting the number of monolayers in the spacer one can tune the dimer to have an exchange coupling ranging from a strong ferromagnetic (at 1-4 ML) to an antiferromagnetic one (5, 7 ML) [1].

We can thus conclude that the exchange coupling of an adatom to a monolayer across a paramagnetic spacer oscillates with the thickness of the latter. This provides reliable means for stabilizing the spin of the adatom in either a ferromagnetic or an antiferromagnetic con-

---

[1] At larger spacer thicknesses it converges to the value of a clean surface.

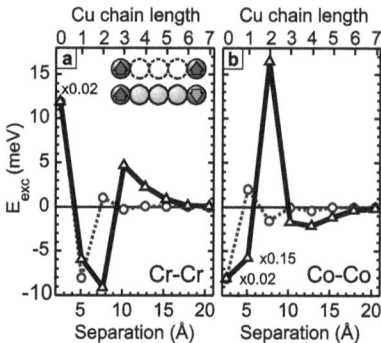

Figure 3.3: (color online). (a) The exchange interaction between two single Cr atoms on a Cu(111) surface (dotted blue line, open circle) and the exchange interaction between those atoms linked with a nonmagnetic Cu chain (solid black line, open triangles). (b) The same dependencies for a Co pair. The exchange energy is calculated via the following formula $E_{exc} = E_{\uparrow\uparrow} - E_{\uparrow\downarrow}$. First several points of the curves were rescaled for clarity. The scaling factors are given next to respective data points.

figuration with respect to the magnetic moment of the monolayer. The competition between interatomic and atom-layer couplings in a dimer allows one, by adjusting the overlayer thickness and the interatomic separation, to tailor the exchange coupling between single magnetic impurities on a surface.

It must be noted here, that a different manifestation of the same physical phenomenon has been described by Uchihashi et al. [157]. In low-temperature scanning tunneling spectroscopy Uchihashi and coworkers have observed that the Kondo temperature T_K of Co atoms adsorbed on Cu/Co/Cu(100) multilayers varies between 60 and 134 K as the Cu film thickness decreases from 20 to 5 atomic layers. The observed change of T_K was attributed to a variation of the density of states at the Fermi level E_F induced by quantum well states confined to the Cu film [157].

3.2 Coupling adjustment by linking with nonmagnetic chains

In the present section we would like to demonstrate that the quantum interference of electrons in quasi-one-dimensional engineered nanostructures, such as the linear atomic chains on a surface [153, 154], can be used to the same end as the quantum confinement in an overlayer.

To demonstrate how artificial nonmagnetic nanostructures introduced in between a pair of magnetic impurities on a Cu(111) surface affect the spin alignment within the dimer we choose, as a model system for our studies, close-packed Cu chains linking a pair of magnetic impurities, following the work of Lagoute et al. [154], who studied close-packed Co-Cu chains of various length and composition, assembled from single Co and Cu atoms on Cu(111) by atom manipulation in a low-temperature scanning tunneling microscope. In our calculations, the total chain length is varying from 2 to 15 atoms. Calculations for a pair of single adatoms

3.2 Coupling adjustment by linking with nonmagnetic chains

were performed for the same positions of magnetic impurities as were used in the case of chains.

Exchange interactions between two single Cr adatoms and in Cr pairs linked with Cu chains are plotted in Fig. 3.3a with the dotted blue line and the solid black line respectively. Nearest neighbor Cr dimer exhibits strong antiferromagnetic coupling. The exchange interaction of a Cr pair at a 5.11 Å separation reverses its sign to ferromagnetic, but its magnitude is strongly reduced because a direct overlap of wave functions of Cr adatoms is not possible. At further separations the exchange interaction oscillating decays to zero. The decay rate at large separations is determined by the two dimensional free-electron-like electron gas stemming from the surface state arising in the projected bulk band gap of Cu(111). It was demonstrated that the exchange interaction in such kind of systems oscillates at large distances with a period determined by the surface state momentum at the Fermi energy [158]. Amplitude of these oscillations decays proportional to r^{-2}, where r is the adatom-adatom separation.

Let us trace now the effect of a Cu chain. The only Cu atom inserted in between Cr dimer at 5.11 Å separation slightly reduces the exchange interaction in comparison to the reference case of single Cr adatoms. But longer mixed chains exhibit drastic differences from the reference case. For instance, two Cu atoms linking the Cr pair reverse the sign of the exchange interaction and strongly enhance its magnitude to 9 meV. The system with three Cu atoms in the chain has the antiferromagnetic exchange interaction equal to 5 meV which is about 10 times larger than the magnitude of ferromagnetic exchange interaction of the reference system. Further elongation of the chain length results in rapid decay of the exchange interaction.

Profound effect of short chains on the exchange interaction and its decay in long ones can be explained by competition of two factors. On the one hand, the density of conduction electrons at Cu atoms is much higher than that of the surface state, so interference effects are expected to be more pronounced and therefore exchange energies can reasonably be higher. On the other hand, each Cu atom scatters conduction electrons into the substrate, so interference patterns should rapidly decay with the increase of the Cu chain length.

Figure 3.3b shows the effect of a Cu chain on the exchange interaction between Co adatoms. A close packed Co dimer on Cu(111) exhibits a strong ferromagnetic coupling with a strength of 0.4 eV. It is evident that short Cu chains significantly affect the exchange interaction as it has just been demonstrated for the case of Cr magnetic impurities. A single Cu atom inserted between Co atoms in a dimer at 5.11 Å separation changes the sign of the exchange interaction and enhances its magnitude. Two Cu atoms permit one to get an antiferromagnetically coupled Co pair. Further increase of the chain length reverses the sign of the exchange interaction back to ferromagnetic. Magnitude of the exchange interaction in presence of a linking chain is of the order higher than that for the reference case. Similar to the case of Cr impurities, magnitude of the exchange interaction between Co coupled through the Cu chain rapidly decay in large chains.

Yet another way to tailor the exchange interaction is to use mixed pairs of magnetic impurities. In Fig. 3.4 we present our results for Cr-Co pairs. Co and Cr in a close packed dimer are coupled antiferromagnetically with an energy of 78 meV. This energy is almost one order of magnitude lower than that for a homonuclear Cr dimer. Non-linked Co and Cr adatoms at 5.11 Å are coupled ferromagnetically with an energy of 5.5 meV. At larger distances their exchange interaction is negligibly small. Short Cu chains inserted between Co

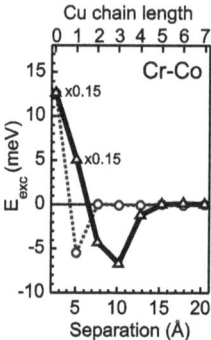

Figure 3.4: (color online). The exchange interaction between Co and Cr adatoms on a Cu(111) surface (dotted blue line, open circle) and the exchange interaction between those atoms linked with nonmagnetic Cu chains (solid black line, open triangles). First several points of the curve were scaled down for clarity. The scaling factors are given next to respective data points.

and Cr significantly enhance the magnitude of the exchange interaction. A single Cu atom placed between Co and Cr reduces the antiferromagnetic exchange interaction to 30 meV. It is important to note that the same structures with pure Co and Cr dimers exhibit exchange interactions of the same order but of the opposite sign. The next points of the graph also demonstrate significant differences from cases of homogeneous magnetic pairs. Magnetic impurities of different species thus provides a wide range possibilities in tailoring the exchange interaction.

To summarize the idea: a modification of the system's magnetic properties can be achieved by introducing nonmagnetic atomic chains in between the adsorbed adatoms. This changes the propagation medium for the conduction electrons, that mediate the exchange coupling in the system. Atomic chains can enhance the exchange interaction between magnetic impurities, can change the sign of the exchange interaction or quench in completely.

3.3 Confinement on islands as a tool for magnetic coupling control

In the present section we discuss the possibility to utilize the quantum confinement of surface state electrons on regular self-assembling structures to tailor the exchange interaction of single adatoms adsorbed on top of them. The main motivation for this particular study was the fact, that the effect of the surface state electrons confinement on the exchange interaction of single adatoms (proven to exist in artificially constructed structures [159, 34]) is hard to exploit in surface engineering, due to the cumbersome procedure involved in constructing such structures (like quantum corrals) in an atom-by-atom way.

Fortunately, there exist natural or self assembling structures which can also be utilized for surface electrons confinement. For instance, it is well known that at certain conditions an epitaxial growth of Ag on Ag(111) [160] and Cu on Cu(111) [161] results in the formation

3.3 Confinement on islands as a tool for magnetic coupling control

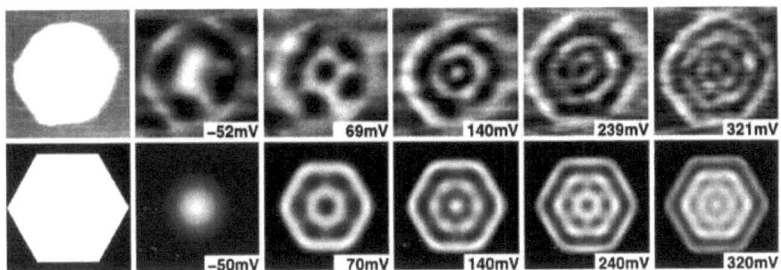

Figure 3.5: Upper row: topographic image of an approximately hexagonal Ag island on Ag(111) (area $\sim 94\ nm^2$), and a series of dI/dV maps recorded at various bias voltages (at $T = 50\ K$). Lower: geometry of a hexagonal box confining a two-dimensional electron gas, and the resulting local density of states. The figure is taken from [162].

of hexagonal islands and vacancy craters.

A quantitative study of the quantum confinement of surface electrons on nanoscale Ag islands on Ag(111) has been carried out by means of scanning tunneling microscopy and spectroscopy (STM/STS) by Li et al. [162, 163]. Such close-packed islands grown on an Ag(111) surface are very stable structures, so standing electron waves can be observed over a wide range of voltages. Differential conductance maps taken above individual islands exhibit strongly voltage-dependent features. Figure 3.5 shows a typical series of differential conductance maps acquired above an island at different bias voltages [162, 163]. Standing wave patterns appear at energies higher than -65 meV (the Ag(111) surface state band bottom energy). Li et al. [162, 163] have identified the standing waves as originating from surface state electrons, confined by the rapidly rising potential at the edges of the island. Simulated LDOS of a 2D electron gas confined to a hexagonal potential well, calculated by the authors [162, 163] for comparison, are demonstrated in the lower row of Fig. 3.5.

These studies have confirmed that the confinement can be observed at all island sizes down to the smallest ones. It has also been proposed [162, 163] that an analogous effect should exist on Cu(111) and Au(111). True enough the confinement in hexagonal craters (vacancy islands) has been recently observed in an STM experiment on Cu(111) [164, 166] (see Fig. 3.6) and the spin-dependent quantum confinement has been extensively studied on Co/Cu(111) nanoislands [159, 165]. Figure 3.7(a and b) shows dI/dV maps taken by Pietzsch et al. [165] on Co/Cu(111) islands at bias voltages allowing to observe the standing wave patterns on both the Cu substrate and on the Co islands. Theoretical calculation results for a similar system by Niebergall et al. are shown in Fig. 3.8. Observed patterns present a vivid example of the efficiency of spin-dependent quantum confinement. As a matter of fact, it was the existence of those fascinating systems that has motivated our study of the interaction of single atoms on hexagonal islands, described in details in the present section.

As a model system for our studies we have chosen pairs of Co atoms adsorbed on top of hexagonal Cu islands of various sizes residing on a Cu(111) surface (Fig. 3.9).

The choice was governed mainly by the fact that such a system is relatively easy to produce experimentally (as discussed in the introductory chapter) yet represents a good model for a system of magnetic adatoms adsorbed on (111) surfaces of noble metals, which

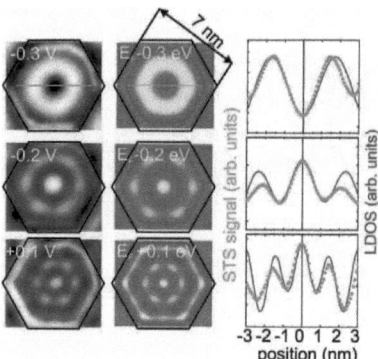

Figure 3.6: STS images of dI/dV at the indicated gap voltage at 1 nA (left column) and the calculated LDOS at various energies of the 7 nm vacancy island. The solid hexagon marks the vacancy island rim. The right column shows horizontal line scans along the center of the vacancy island, as indicated by the green line for different energies. Solid line: calculation (center column), symbols: experimental data (left column). Figure from [164].

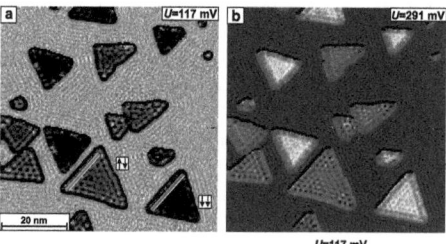

Figure 3.7: Co islands on a Cu(111) surface. dI/dV maps at sample bias voltages as indicated in the figure, showing standing wave patterns on both Cu substrate and Co islands. Arrows indicate islands being magnetized parallel (↑↑) or antiparallel (↓↓) to the tip magnetization. The figure is taken from [165].

3.3 Confinement on islands as a tool for magnetic coupling control

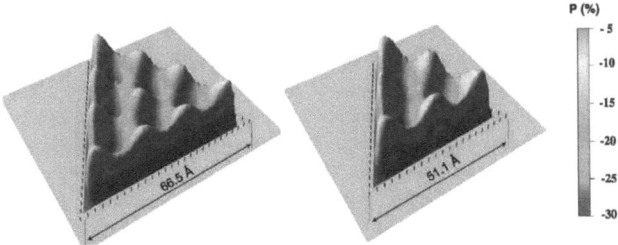

Figure 3.8: The spin polarization of surface-state electrons on triangular Co islands on Cu(111); calculations are performed for $E = 0.5$ eV above the Fermi level. Adopted from [159].

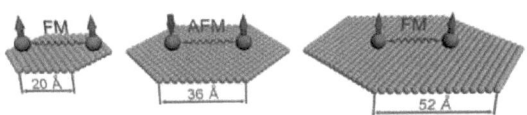

Figure 3.9: (color online) System setup used in our calculations to demonstrate the possibility of tailoring the exchange coupling between single adatoms adsorbed on islands by adjusting the island's size. The separation of adatoms remains constant.

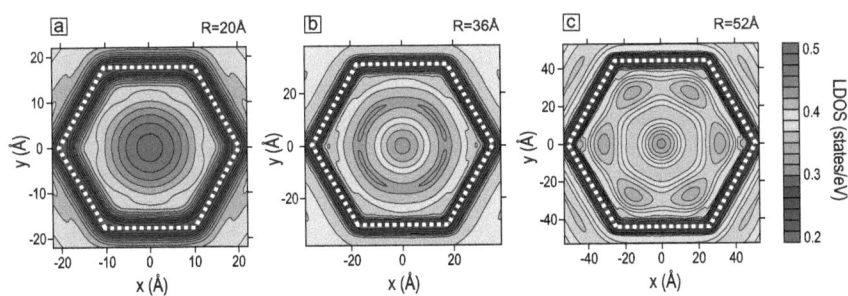

Figure 3.10: LDOS distribution on islands with R=20 (a), 36 (b) and 52 Å (c).

increases the generality of attained conclusions. To study the influence of the islands size on the electron confinement and the exchange coupling of adsorbed Co atoms we have selected hexagonal islands with circumscribed circle radii of 20 Å, 36 Å and 52 Å (side lengths of 8, 14 and 20 atoms respectively). To investigate the quantum confinement induced by islands we calculated the LDOS in vacuum some 2 Å above clean Cu islands. To produce LDOS maps the integrated density of electronic states at the Fermi energy in each atomic sphere of the first vacuum layer above the island has been calculated. For a clearer picture the values have then been interpolated to produce the $\eta = \eta(x, y)$ (where x and y are the in-plane coordinates and η is the local density of states) contour map shown in Fig. 3.10(a, b and c) for 20 Å, 36 Å and 52 Å islands respectively. The confining factor for the electrons in case of an island is the vacuum barrier existing at it's boundaries. Thus the confinement region inherits the shape of the island's boundary and an electron confined in it behaves like a particle in a hexagonal box which clarifies the LDOS distributions shown in Fig. 3.10(a-c) [162, 163]. Those distributions closely resemble the first three eigenmodes of a particle with a wavelength of about 30 Å which corresponds to the Fermi wavelength of Cu surface state electrons. The first and the third distributions display a density maximum in the center of the hexagon with an additional maximum at the boundary on the 52 Å island. The second mode has its high density region in the form of a ring with a diameter of about $30-35$ Å and depletion zones both at the center and at the boundaries. The confinement causes the 20 Å island to acquire the highest (among the three) absolute local density of electronic states at the center, while the LDOS map of the 52 Å island displays, as can be expected, the most profound hexagonal features. The obtained distributions are very similar to those observed in hexagonal vacancy holes on Cu(111) [164] (Fig. 3.6) which have been comprehensively analyzed in Ref. [166].

Considering that surface electrons act as mediators of the indirect exchange interaction it is most likely that a change of the LDOS up to 30% (0.5 st/eV in the center of the 20 Å island versus 0.35 st/eV on the the 35 Å one) will lead to a significant modulation of the exchange coupling of magnetic atoms adsorbed on top. However, it should be noted here, that besides the intrinsic electron density redistribution the coupling constant is effectively determined by the phase relation of surface state electrons scattered at both impurities which is in turn profoundly affected by the introduction of reflective vacuum barriers at the boundary of the island. By changing the island's size we alter both the intrinsic density of surface state electrons and the scattering geometry. However, although the results of *ab initio* calculations incorporate both effects, it is virtually impossible to separate those two contributions.

To study the cumulative impact of the quantum confinement of surface electrons on the exchange interaction of magnetic atoms adsorbed on top of an island we have calculated the exchange coupling energies between single Co atoms residing on hexagonal Cu islands of the three above-mentioned sizes. In regard of the fact that the placement of two adatoms on a nanometer-scale island allows for a vast amount of variants let us consider first one of the most logical choices, namely the case when one of the adatoms is placed in the center of the island and the other occupies one of the remaining adsorption sites. The resulting exchange energy landscape $E_{exc}(x, y) = E_{exc}(x_1 = 0, y_1 = 0, x_2 = x, y_2 = y)$ (where (x_1, y_1) and (x_2, y_2) are the in-plane coordinates of the first and the second Co adatoms) for the 20 Å island is given in Fig. 3.11. It must be noted that the coordinate zero does not precisely coincide with the geometrical center of the island but rather with an adsorption site closest

3.3 Confinement on islands as a tool for magnetic coupling control 69

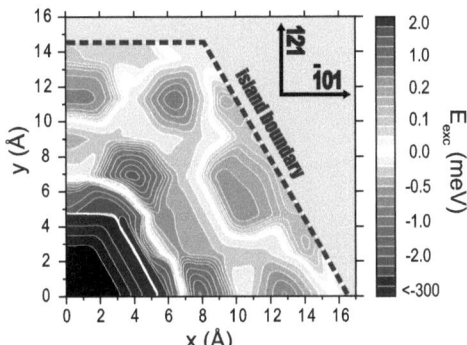

Figure 3.11: (color) A map of the exchange interaction energy of two Co adatoms adsorbed on a hexagonal Cu island of 20 Å in radius, given that one of the adatoms is residing in the island's center.

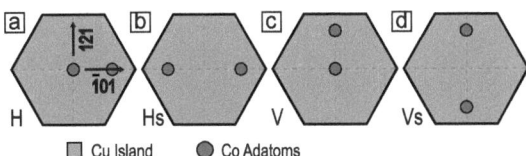

Figure 3.12: (color online) A scheme of principal arrangements of Co adatoms on an island chosen for investigation: (a) "horizontal" (H), (b) "horizontal symmetric" (Hs), (c) "vertical" (V) and (d) "vertical symmetric" (Vs).

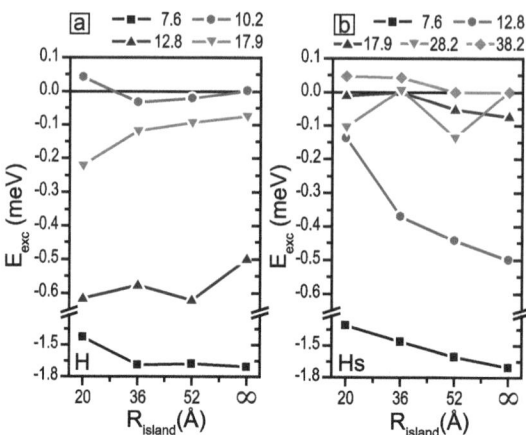

Figure 3.13: (color online) (a) Exchange coupling energies of two Co adatoms aligned according to configuration "H" (see Fig. 3.12) at separations of 7.6 Å (black squares), 10.2 Å (red/dark-grey circles), 12.8 Å (blue/grey triangles pointing up) and 17.9 Å (green/light-grey triangles pointing down) as a function of the island's size R_{island}. (b) Exchange coupling energies of two Co adatoms aligned according to configuration "Hs" (see Fig. 3.12) at separations of 7.6 Å (black squares), 12.8 Å (red circles/dark-grey), 17.9 Å (blue/grey triangles pointing up), 28.2 Å (green/light-grey triangles pointing down) and 38.2 Å (cyan/light-grey diamonds) as a function of the island's size. In both cases corresponding exchange energy values on a clean surface are given as a last point of each curve (denoted on the horizontal axis as ∞). Connecting lines are only there to guide the eye.

to it which is situated some 1.5 Å away. It can be seen that starting from intermediate separations of ∼8 Å the map exhibits two principal directions that differ significantly: along the $[\bar{1}01]$ and the $[1\bar{2}1]$ vectors of the surface. Those directions are dictated by the six-fold symmetry of the islands. Most intermediate directions reveal an exchange behavior closely resembling that along one of the principal directions. Relying on this fact we have chosen to limit our investigations to four main arrangements aligned along those two main direction. For each direction the two adatoms were either aligned symmetrically with respect to the center of the hexagon or one of them was placed at the center and the other at various separations along the respective direction. A sketch of all four configurations is given in Fig. 3.12. For ease of notation let us designate those configurations as "horizontal" (H), "horizontal symmetric" (Hs), "vertical" (V) and "vertical symmetric" (Vs) (Fig. 3.12(a,b,c and d respectively)).

Let us now take a closer look at how the quantum confinement on islands influences the exchange interaction. Fig. 3.13a shows the exchange energies of two Co adatoms adsorbed in an "H" (see Fig. 3.12) configuration at various separations as a function of the island's size R_{island}. For comparison exchange energies of the same adatoms at the same separation on a clean Cu(111) surface are given as the last point of each curve marked on the R_{island}

3.3 Confinement on islands as a tool for magnetic coupling control

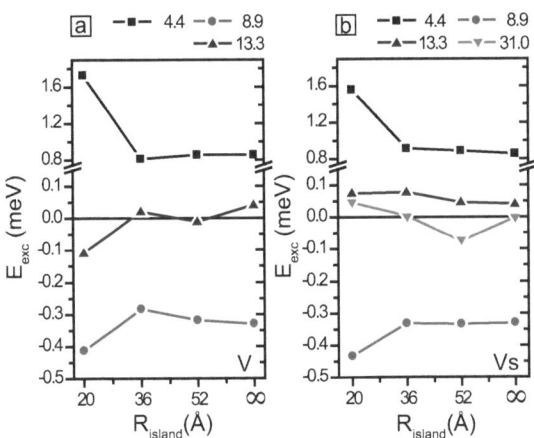

Figure 3.14: (color online) (a) Exchange coupling energies of two Co adatoms aligned according to configuration "V" (see Fig. 3.12) at separations of 4.4 Å (black squares), 8.9 Å (red/dark-grey circles) and 13.3 Å (blue/grey triangles pointing up) as a function of the island's size R_{island}. (b) Exchange coupling energies of two Co adatoms aligned according to configuration "Vs" (see Fig. 3.12) at separations of 4.4 Å (black squares), 8.9 Å (red/dark-grey circles), 13.3 Å (blue/grey triangles pointing up) and 28.2 Å (green/light-grey triangles pointing down) as a function of the island's size. In both cases corresponding exchange energy values on a clean surface are given as a last point of each curve (denoted on the horizontal axis as ∞). Connecting lines are meant as a guidance for the eye only.

axis by ∞. The figure indicates that at smaller separations of 7.6 Å (black squares) the exchange energy is gradually reduced (increasing the ferromagnetic (FM) coupling) with increasing island's size in the range of 20 − 25% of the clean surface value. At intermediate separations of 10.2 Å and 12.8 Å (red circles and blue triangles pointing up respectively) the dependence becomes irregular in its behavior allowing either for a switching (at 10.2 Å) or for a 20% increase (at 12.8 Å) of the exchange coupling between the adsorbates. At larger separations of 17.9 Å (green triangles pointing down) the quantum confinement allows one to increase the FM coupling energy to up to twice its value on a clean surface. Similar exchange dependencies for the "Hs" configuration are presented in Fig. 3.13b. Here as well, at small and intermediate separations of 7.6 Å (black squares) 12.8 Å (red circles) and 17.9 Å (blue triangles pointing up) the exchange energy is reduced by the confinement decreasing the coupling. But an even more significant effect can be observed at large separations of 28.2 Å and 38.4 Å (green triangles pointing down and cyan diamonds respectively). At a clean surface the exchange coupling at such separations is negligibly small. The quantum confinement on an island, however, allows one through reflection and focusing of scattered electrons by the island's boundaries to restore the coupling to values reaching up to over 100 μeV.

To generalize the picture even further let us take a look at both the "V" and "Vs" configurations. The exchange coupling dependencies for those configurations are presented in Fig. 3.14(a and b respectively). Though different values of interatomic separations superimposed by the *fcc* lattice do not allow us to make a one to one comparison with "H" and "Hs" cases the dependencies show a rather similar behavior and tendencies. Once again at small separation a deliberate choice of the island size can allow us to either reduce or increase the interatomic coupling between the adsorbates. At intermediate separations we yet again have the possibility to switch the exchange coupling by adjusting the island size. And at larger separations of about 30 Å the initially small coupling at a clean surface can be restored to palpable values.

We can thus conclude that the quantum confinement of surface electrons to nanoislands is a suitable tool for tailoring the exchange interaction of magnetic adsorbates. In some systems (for instance Cu/Cu(111) or Ag/Ag(111)) such islands can be relatively easily obtained by epitaxial growth and their size distribution can be controlled by growth conditions. The dependence of the surface electron density distribution and the phase relations of scattered electrons on the island's size makes the latter a convenient adjustment parameter. Besides the island size, the alignment of magnetic adsorbates with respect to the island has a notable effect on the interatomic exchange interaction and thus can be varied to achieve a desired effect. This fact might make the quantum confinement on islands a good candidate for future spintronic applications.

Concluding the chapter we would like to once again, daring to repeat ourselves, underline the fact, that quantum confinement of electrons at surfaces can prove to be an extremely powerful tool for tailoring the magnetic interaction between various surface-based structures.

Chapter 4
Probing the exchange interaction

Despite the considerable experimental power of the magnetic interaction probing techniques presented in Chapter 1, the well of techniques for probing the interatomic exchange coupling is far from being drained. In the present chapter we tackle the problem from the new point of view. We analyze the effect of the exchange interaction on the spin-dependent localization of the surface state.

It is well known, that (111) surfaces of some noble metals produce an electronic surface state, which in it's properties resemble a free 2-dimensional electron gas (2DEG). The origin of the surface state is the trapping of electrons between the vacuum barrier and the band gap of the metal bulk.

An impurity immersed in such a 2DEG presents an additional potential for the surface state electrons. The theorem of Simon [168] predicts that a Hamiltonian within an arbitrary asymptotically attractive 2D potential has a single bound state. It has been shown [169, 167, 170], that the localization of a 2D Shockley surface state on an impurity potential manifests itself as a split-off bound state just below the surface state band bottom. dI/dV spectra measured by Limot and coworkers above Ag(111) and Cu(111) surfaces and above single Co atoms adsorbed on those surfaces are presented in Fig. 4.1 (a,c) and (b,d) correspondingly. The localization seen in the lower panels just below the surface state band bottom (shown in upper panels) was ascribed by the authors [167] to the Simon localization mechanism [168]. Similar states have been predicted to exist at nonmagnetic Cu chains [171].

To trace the effect of the exchange interaction on the bound state, we consider single Co adatoms and atomic pairs at different separations adsorbed on a Cu(111) surface. A clean Cu(111) surface has an intrinsic surface state of Shokley type. It manifests itself as an onset in the density of states at about -0.5 eV (below the Fermi energy). The density of states in vacuum above a clean Cu(111) surface is plotted in Fig. 4.2c as a gray curve with a filled area below. If we now adsorb a single Co adatom on the surface, the surface state becomes localized by the attractive potential of the impurity. The result of the LDOS calculation above a single Co adatom is presented in Fig. 4.2c for majority (red solid) and minority (blue dashed) electrons. The localization of the surface state has lead to the formation of a narrow peak in both spin-up and spin-down channels at about -0.57 eV, which is just below the surface state band bottom. Due to different scattering properties of the magnetic adatom for the spin-up and spin-down surface electrons, the bound state acquires a certain spin polarization. The position, the shape and the spin splitting of the bound state peak are in good agreement with the results obtained experimentally and theoretically by other groups [169, 167, 172, 173]. To be consequent, let us note, that the majority peaks around

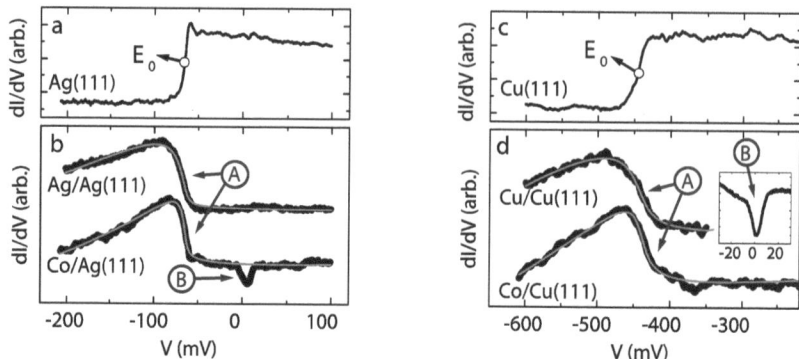

Figure 4.1: dI/dV spectrum over (a) the center of a 20×20 nm^2 defect- and impurity-free area of Ag(111) and (b) the center of a Ag and of a Co atom on Ag(111) (feedback loop opened at 200 MΩ; spectra are shifted vertically for clarity). Light-colored solid lines: fits described in [167]. (c and d) similar graphics for the Cu(111) surface with Cu an Co adatoms adsorbed on top (feedback loop opened at 100 MΩ). Figures taken from [167].

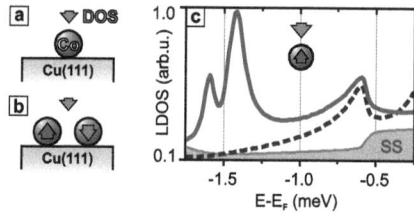

Figure 4.2: The geometry considered in calculations: an adatom (a) and a dimer (b) of Co adsorbed on a Cu(111) surface. The majority (red solid) and minority (blue dashed) LDOS above a single Co adatom (c). The gray filled curve represents a surface state of a clean Cu(111) surface.

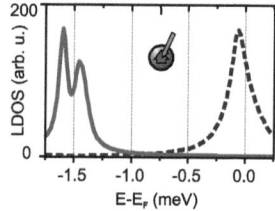

Figure 4.3: The majority (red solid) and minority (blue dashed) LDOS at a single Co adatom. The peaks in both spin channels are mainly of a d character.

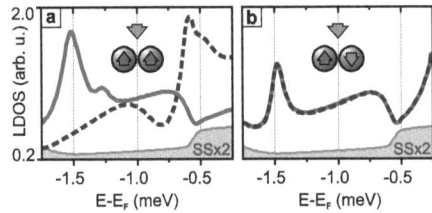

Figure 4.4: The majority (red solid) and minority (blue dashed) LDOS above the middle of a compact Co dimer ($d = 2.55$ Å), aligned along the $[\bar{1}01]$ direction of a Cu(111) surface, in a ferromagnetic (a) and an antiferromagnetic (b) configurations. The gray curve shows the surface state above a clean Cu(111) surface. It has been scaled by a factor of 2 for clarity.

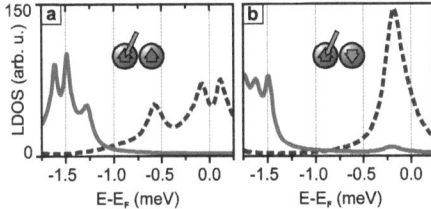

Figure 4.5: The majority (red solid) and minority (blue dashed) LDOS at one of the atoms in a compact Co dimer ($d = 2.55$ Å), aligned along the $[\bar{1}01]$ direction of a Cu(111) surface, in a ferromagnetic (a) and an antiferromagnetic (b) configurations. The peaks in the LDOS are mainly of the d character.

-1.5 eV can be ascribed to the hybridization of the surface electrons with the d orbitals of the Co adatom (the LDOS at the adatom is shown in Fig. 4.3).

If another Co atom is introduced to the vicinity of the first one, the manner in which each of the adatoms interacts with the surrounding electrons becomes affected by the interatomic interaction. In this article we discuss only the magnetic exchange interaction between the adatoms, as it is the magnetic coupling that is of the greatest interests for potential spintronic applications. At small interatomic separations (< 3 Å) the main source of exchange interaction between adatoms is the direct overlap of atomic orbitals. At larger distances the magnetic interaction is mainly mediated by conduction electrons of the substrate (RKKY mechanism). It was shown [158], that the RKKY nature of the interaction causes the exchange coupling to oscillate with a wavelength determined by the copper surface state Fermi wave vector and decay quadratically with the distance. The fact, that the system is reasonably simple to produce and control experimentally and assumes a wide range of exchange coupling strength values, makes such a system an ideal testing ground for both theoretical and experimental exchange interaction probing methods.

First of all, let us consider two Co atoms adsorbed on fcc hollow sites along the $[\bar{1}01]$ direction of a Cu(111) surface, forming a compact dimer. The LDOS above the middle of the dimer is shown in Fig. 4.4 for a (a) ferromagnetic (FM) and an (b) antiferromagnetic (AFM) alignment of spins. The energetically stable configuration is the FM one. The total energy calculations yield an exchange coupling energy of -432 meV. Let us analyze the origin of various peaks in the LDOS. If we consider the LDOS at one of the atoms of the dimer (Fig. 4.5) in FM (a) and AFM (b) configurations. It immediately becomes evident, that majority peaks around -1.5 eV and minority peaks around the surface state band bottom (-0.5 eV) in the FM case are of atomic origin, i.e. arise from the hybridization of s surface electrons with majority and minority d states of Co respectively. The remaining peaks (around -1.1 eV for minority and -0.8 eV for majority electrons) stem from the formation of the bound state. If we consider the position of the bound state peaks it can be seen, that the potential of the FM dimer has caused the bound state to acquire a larger splitting than in the case of a single adatom and become broadened. At the same time the LDOS for an AFM configuration displays no spin-splitting at all.

This phenomenon can be easily understood if we consider the system in the framework of a two-state spin model. According to this model, one should expect, that the interatomic interaction will cause localized electronic states to become split into a bonding-antibonding doublet in both the spin-up and the spin-down channels. However, the intrinsic width of the bound state peak does not allow us to resolve the split resonances and thus we only observe a broadening of the localization peak. Moreover, the spin degeneracy in the AFM case is also a logical consequence of the theory. A similar idea and explanation were proposed for impurities on metal surfaces [174] and in semiconductors [175] and have been experimentally proven to be quite feasible [176].

Another way to understand the results of the calculations is to consider the dimer as a single scattering entity with a net spin being the sum of individual atom's spins. Then the AFM dimer can be regarded as a nonmagnetic scattering entity which immediately explains the spin-degeneracy of the bound state. It also explains the increased spin-splitting of the bound state arising at a FM dimer as compared to a bound state above a single adatom, as the net spin moment of the dimer is about twice the moment of a single magnetic impurity. Another thing that has to be mentioned in the context of a compact magnetic dimer is that the coupling in this case is governed by the direct overlap of the atomic wave functions; and thus our reasoning might be slightly impaired by the fact that some magnetic dimers are known to have a noncollinear ground state [177, 178], in which case the two-state model for atomic spins is no more applicable. However this does not hold for larger separations and therefore does not affect the reasoning in general.

Let us now trace the changes in the surface state localization as we gradually increase the distance between the Co adatoms. For each of the interatomic separations ranging from 2.55 to 15.33 Å we determined the systems ground state and the corresponding exchange coupling energy (left panel of Fig. 4.6). Then we calculated the LDOS above a point midway between the adatoms (right panel of Fig. 4.6) for the system in its ground state. As was already mentioned, the ground state for a compact dimer is a FM one. At the second nearest neighbor separation the Co pair in the ground state displays an AFM alignment of spins. The corresponding LDOS shows, as expected, no spin splitting. The width of the bound state peak is decreased and its maximum is shifted to higher energies. At 7.66 Å separation (3rd nearest neighbors) the spin-polarization of the ground state LDOS is recovered and the corresponding spin coupling is a FM one. Both peaks are positioned at higher (with respect to the second nearest neighbor) energies. By further increasing the separation we acquire a sequence of LDOS-es and exchange coupling energies. As in the first three cases, it can be clearly seen that for all configurations with an AFM ground state the bound state is spin-degenerate, while for FM configurations it displays a splitting, that decreases with increasing atom-atom separation. Thus the spin-splitting of the bound state can serve as an

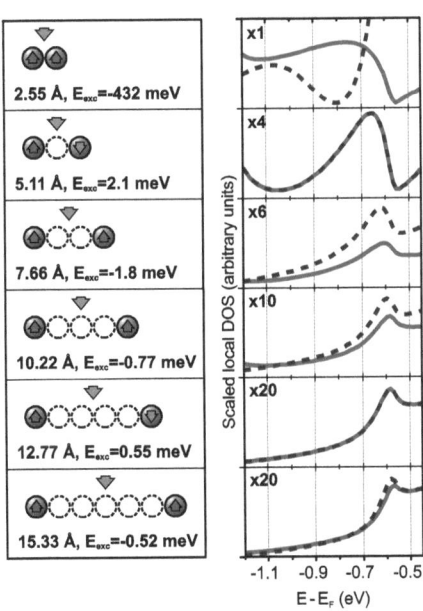

Figure 4.6: The majority (red solid) and minority (blue dashed) LDOS above the middle of Co dimers at different separations (right panel) and the corresponding configurations, interatomic separations and exchange coupling energies (left panel). LDOS have been scaled for clarity. The scaling factors are given near each curve.

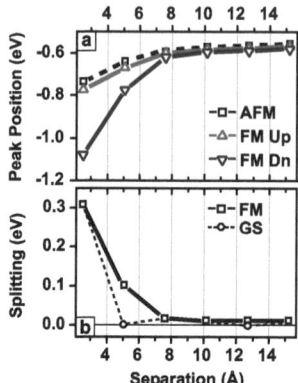

Figure 4.7: The position of the bound state of Co dimers as a function of the interatomic separation for antiferromagnetic (dashed line, squares) and ferromagnetic (spin-up – solid red curve, triangles pointing up, spin-down – solid blue, triangles pointing down) configurations (a). The amount of spin-splitting of the bound state above a ferromagnetically coupled dimer (solid curve, squares) and a dimer in the ground state (GS) configuration (dashed, circles) versus the interatomic distance (b)

indicator of the exchange coupling sign, being zero for antiferromagnetic alignment of spins and non-zero for the ferromagnetic one.

To get a grasp on the quantitative dependance of the bound state peak position on the exchange coupling strength, we present the bound state energies in Fig. 4.7a as a function of the distance between Co adatoms. The black dashed curve shows the position of the bound state for AFM aligned dimers. With increasing interaction at smaller separations the peak gets shifted to lower energies. In the case of a ferromagnetic configuration, the spin-up bound state (solid red curve, triangles pointing up) displays a similar behavior. However, the spin-down bound state (solid blue, triangles pointing down) occupies a lower energy at all interatomic distances. Moreover, the amount of splitting also increases with the interaction strength. The values of the splitting for the ferromagnetic configuration are given in Fig. 4.7b by a solid curve and square markers. The dashed curve represents the spin-splitting in the ground state configuration. The shift of the bound state peak position (with respect to the position of the bound state above a single adatom) shows a monotonous decrease with increasing interatomic separation. This is valid for both FM and AFM configurations. However, in absence of a reference point the position of the localization peak alone can not be regarded as a reliable measure of the exchange coupling. The spin-splitting of the bound states provides a more reliable source of information. One could expect, that the splitting amount would decay with interatomic distance, following a power-law, as is usual for values describing conduction-electrons mediated processes; however, the investigated distances are not large enough and the amount of splitting is too small to make any conclusive statement. Still, it is clear that both the position of the bound state and the amount of its splitting provide a valuable source of information about the magnetic exchange interaction between single adatoms on a surface.

The application of this approach is by no means restricted to particular adatom species.

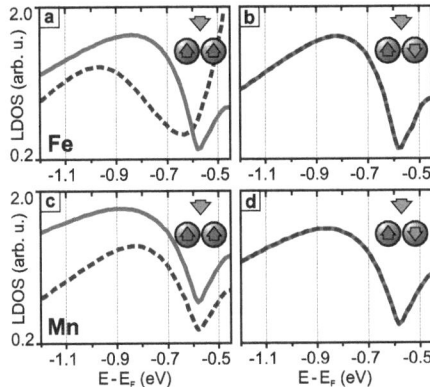

Figure 4.8: The majority (red solid) and minority (blue dashed) LDOS above the middle of compact Fe (a,b) and Mn (c,d) dimers ($d = 2.55$ Å), aligned along the $[\bar{1}01]$ direction of a Cu(111) surface, in a ferromagnetic (a,c) and an antiferromagnetic (b,d) configurations.

Our calculations show, that the bound state peak demonstrates the same behavior for most of 3d adatoms. For example, LDOS above the middle of compact Fe and Mn dimers is shown in Fig. 4.8. We also assume, that this approach can be generally applied to any surface possessing a surface state and any kind of magnetic dimers.

Another aspect worth mentioning, is the stability of the approach with respect to the location of the point of measurements/calculation, as this might be relevant for its experimental application. The deviation from the axes of symmetry of the dimer causes the LDOS of an AFM configuration to start loosing the spin-degeneracy. However, due the s character of the localized electrons the susceptibility of the LDOS to the symmetry breaking is only moderate in the AFM case and hardly noticeable in the FM case. Therefore, at small and intermediate separations, which are the relevant ones for technological applications, the proposed approach should be stable against small variations of the point of measurement and thus retain it's validity.

In conclusion of the chapter, we have shown, that the localization of the surface state is strongly affected by the interaction of atomic spins and can possibly be utilized to determine the exchange coupling between single impurities adsorbed on metallic surfaces. The presence (or absence) of a spin-splitting of the bound state indicates a ferromagnetic (respectively antiferromagnetic) alignment of spins in the system. Moreover, the amount of splitting can be regarded as an indicator of the exchange coupling's strength.

This approach may provide suitable means of exchange coupling probing at small and intermediate separations for future spintronic applications.

Chapter 5

Magnetism of buried nanostructures

This chapter is a logical extension of the previous one to a new and profoundly interesting class of objects: single atoms and small clusters buried *beneath* a metallic surface. First of all we face the question whether it is at all possible to sense the magnetism of single buried structures on the nanoscale with modern local probe techniques. Having positively answered this question we proceed to probing the exchange coupling between single buried structures and show, that the coupling of buried structures can be deduced from the spin polarization of surface electrons above the structure's burying site.

5.1 Probing the electronic and magnetic properties of buried nanostructures

The advances of the last three decades in the field of local probe methods, such as the scanning tunneling (STM) or atomic force (AFM) [179, 43] microscopies, made it possible to address structural, electronic and magnetic properties of surface structures which have been previously inaccessible. Yet, despite the aptitude of experimental techniques, some of the surface systems remain largely unexplored. Most of the recent attention of the STM and AFM communities has been directed toward nanostructures adsorbed on top of metallic, oxidic or semiconductor surfaces. However, the prospect of using subsurface (embedded) nanostructures as the base for hypothetical future applications (such as the almost proverbial spintronic devices) might be even more promising (see, e.g. [180]). So it is obvious that learning to probe and tailor buried magnetic structures might be a challenge well worth taking.

The question, whether nanostructures buried several monolayers deep inside a surface can be investigated at an atomic scale with a local probe technique has been raised several times in the past two decades. The first definitive positive answer was given theoretically by Crampin [181]; and experimentally by Heinze *et al.* [182] for Ir impurities in metallic surfaces and by van der Wielen *et al.* [183] for Si dopants in semiconductor alloys. The latter relied on the presence of the Friedel oscillations induced by impurities at the surface. The former used a more local mapping of the surface combined with an extensive theoretical (first-principles calculations) support. Following those "proof of principle" experiments several studies have been undertaken to further explore the possibility both experimentally and theoretically [14, 184, 185, 186] and their findings have been extensively utilized in a subsequent series of studies aimed at the determination of electronic structure and position of subsurface

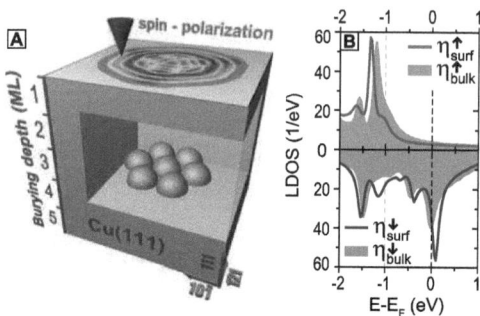

Figure 5.1: (A) The sketch of the studied system - a hexagonal Co cluster of 7 atoms (H7) buried in a Cu(111) surface. (B) Majority (light-red filled area) and minority (light-blue filled area) LDOS in of the central atom of the cluster embedded in Cu bulk. Majority (red solid curve) and minority (blue solid curve) LDOS of the central atom of the cluster embedded into the topmost layer of a Cu(111) surface.

impurities [187, 188] and at studying buried interfaces and lattices [189].

However, at least one aspect remained largely unexplored, namely the possibility to address the magnetism of surface-embedded nanostructures. Yet precisely this aspect might be of essence for possible future applications. This realization might have been the motivation to reopen the subject of buried nanostructures in the magnetic perspective. So, for example, Weismann et al. [190], in their study of the possibility to probe the topography of the host Fermi-surface utilizing magnetic subsurface point defects (Co atoms), propose, in a manner of speculation, that buried magnetic impurities can be put to work as "nano-sonars" for probing geometric and electronic properties of buried interfaces. They indicate that an extended ferromagnetic nanostructure can play the role of a "spin filter" which splits non-spin-polarized current. Besides they suggest that subsurface defects might be used for direction-specific control over interatomic interactions at the surface (see Fig. 4 (A, B and C) of [190]). In their experimental endeavors they utilize the basic ideas which have already been rather extensively discussed in recent years by Avotina et al. for both para- [187, 191] and ferromagnetic [192] point defects buried in a metallic host. Quite recently the same group (Avotina et al.) has published a comprehensive experiment-oriented theoretical study, addressing the influence of a single magnetic defect or cluster in a nonmagnetic host (metal surface) on the properties of the spin and charge currents in the proximity of a ferromagnetic STM tip [193].

5.1.1 Probing magnetic subsurface impurities with the local density of states at the surface

Here we would like to specifically concentrate on the topic of probing the buried structure's magnetism. Let us first say a few words about the calculational setup which was used to study buried magnetic structures. For calculational studies the choice of a good model system is essential. A prototypical system for studying magnetism at metallic surfaces is Co/Cu(111). It follows most general trends for this class of systems and is very suitable for DFT calculations, as it is known to yield a good agreement with experiments at a relatively

5.1 Probing the electronic and magnetic properties of buried nanostructures

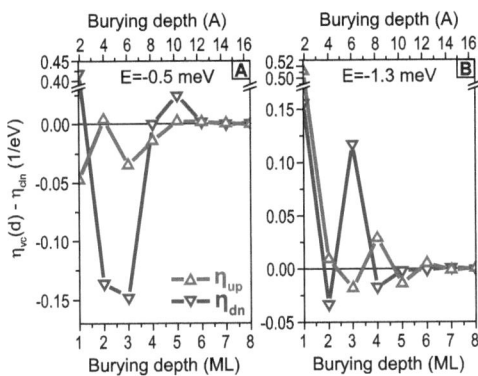

Figure 5.2: Majority (red triangles up) and minority (blue triangles down) LDOS at -0.5 (A) and -1.3 (B) eV in vacuum above the embedding site of an H7 Co cluster versus the burying depth. The LDOS of a host surface has been subtracted from all the curves for clarity.

low computational cost. As for the shape of the nanostructure, the simplest choice would be a cluster consisting of atoms arranged in a regular manner (e.g. a hexagon) all residing in the same layer beneath surface (a flat hexagonal cluster). Such systems can be reproduced experimentally using the self-assembly (or the buffer layer assisted growth) in conjunction with a capping layer deposition. The feasibility of such a technique has been shown, e.g., by Torija et al. [194]. A sketch of the system is presented in Fig. 5.1A. The electronic structure of such a cluster would remain largely unchanged if it is submerged into the surface. A comparison of the local density of states (LDOS) of the central atom of a hexagonal Co cluster, consisting of 7 atoms (H7), is presented in Fig. 5.1B. The LDOS of a cluster residing in the topmost layer of the surface (a red solid curve for majority LDOS (top panel) and a blue solid curve for minority LDOS (bottom panel)) differs from the LDOS of a cluster embedded in the bulk (filled curves) only through a slight change of peak positions and peak intensity distributions, sharing most of the other main features. Several most prominent peaks can be pointed out: at -1.45, -0.44, 0.0 eV for minority and -1.6, -1.2 eV for majority electrons.

In a real experiment, however, the electronic structure of a buried impurity is not a directly accessible value. What modern local probe techniques can access is the LDOS at the surface. Thus it is essential to understand how a buried nanostructure can affect the electron density at the surface above it's burying site. Fig. 5.2 shows the evolution of the LDOS at two selected energies ((A) -1.3 eV and (B) -0.5 eV, each corresponding to a region containing a prominent peak in either the majority or the minority LDOS of the cluster) as the nanostructure is submerged ever deeper into surface. The investigated burying depths range between 1 (surface layer) and 8 monolayers (ML) which corresponds to about $2-17$ Å. The majority LDOS is plotted in red triangles, pointing up, and the minority LDOS - in blue triangles, pointing down. For convenience, the LDOS value of a clean Cu(111) surface at corresponding energies has been subtracted from the curves, so that the presented values would reflect the partial influence of the submerged impurity on the electronic density at the surface. The most remarkable feature is that both the majority and the minority LDOS display an oscillatory behavior. Such behavior is nowadays clearly understood and can be ascribed, similar to [187], to the quantum interference of the s-like states in the paramagnetic spacer between the nanostructure and the vacuum barrier at the surface. The boundary conditions, determining the density of states at the surface, thus depend on the scattering properties of the nanostructure, the vacuum potential, the k-vector of the energy state in question and, of course, on the spacer thickness or the burying depth. Thus any change in the burying depth of the nanostructure would be immediately reflected in the LDOS at the surface, which immediately suggests it as a tool for probing the structure's vertical position.

Furthermore, such oscillatory behavior is not energy-bound to the two selected values presented here. Fig. 5.3 clearly illustrates this point, showing the energy-resolved dependence of the LDOS above the center of the cluster on its burying depth. Pronounce oscillatory behavior can be registered throughout all the energy spectrum in both the spin-up (A) and spin-down (B) channels. It is also evident, that the LDOS at the surface mimics that of the central atom of the buried cluster (Fig. 5.1B) thus confirming the fact, that the spin polarized electronic structure of a buried impurity can be probed by local probe techniques at the surface.

Considering now that the scattering properties of a magnetic nanostructure are spin-dependent it is not hard to justify the differences between spin-up and spin-down LDOS at the surface. This effectively means a presence of a net polarization of surface electrons throughout the LDOS spectrum. As an example Fig. 5.3C displays the spin polarization at the surface $P(E,d)$ as a function of energy and burying depth:

$$P(E,d) = \frac{n_{up}(E,d) - n_{dn}(E,d)}{n_{up}(E,d) + n_{dn}(E,d)},$$

where $n_{up,dn}(E,d)$ is the burying depth and energy resolved density of states for majority and minority electrons at the surface. Clearly the oscillations of the LDOS with increasing burying depth cause the polarization to oscillate accordingly. Both large positive and large negative values of the polarization, ranging between +35 and −50 %, can be observed. This oscillations can be traced up to the burying depths of at least 8 ML (∼17 Å). Similar effects can be observed already for a single buried atom. The corresponding spin resolved LDOS above the burying site and the polarization of surface electrons are presented (similarly to Fig. 5.3) in Fig. 5.4A-C, correspondingly. One can observe the same burying-depth-dependent oscillatory features in both, the LDOS and the polarization, yet their amplitude is considerably less that in the case of a 7-atomic cluster for obvious reasons. Thus it is clear that the polarization along with the LDOS is a very sensitive tool for studying embedded magnetic nanostructures.

However, there are several major questions that still require to be answered. One of them is the question of geometry dependence. How would the polarization at the surface change if we alter the geometry of the cluster, for example by increasing its size. As the simplest comparison one can take clusters of two different sizes (a 7-atomic (H7) and a 19-atomic (H19) hexagonal ones). The polarization above buried H7 (black rectangles) and H19 (red circles) clusters as a function of burying depth at two chosen energies mentioned above (−0.5 eV (A) and −1.3 eV (B), marked with dashed lines in Fig. 5.3) is presented in Fig. 5.5. As an asymptotic case another nanostructure is taken for comparison, namely a whole Co monolayer (ML) embedded into the surface at the same depths as the clusters. The polarization above the monolayer is given in the figure by blue triangles. One might notice that the depths can be logically divided into two regions. When the distance to the surface is

5.1 Probing the electronic and magnetic properties of buried nanostructures 85

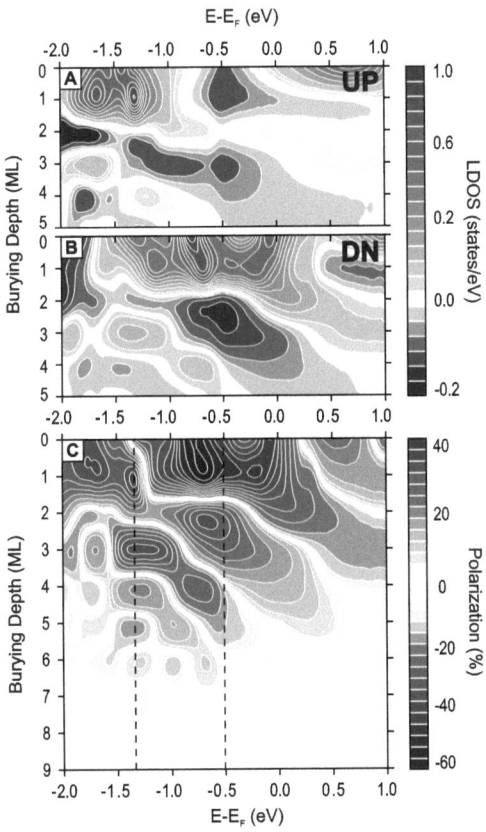

Figure 5.3: Energy resolved majority (A) and minority (B) LDOS (color-coded) in vacuum above the embedding site of an H7 Co cluster versus the burying depth. The LDOS of a host surface has been subtracted for clarity. (C) Polarization (color-coded) $P = P(E,d)$ above a buried hexagonal 7-atomic Co cluster as a function of electron energy and burying depth. The $P(E)$ distributions have been calculated for integer layer numbers and then interpolated for clarity. The dashed vertical lines mark the energies chosen for comparison in Fig. 5.2.

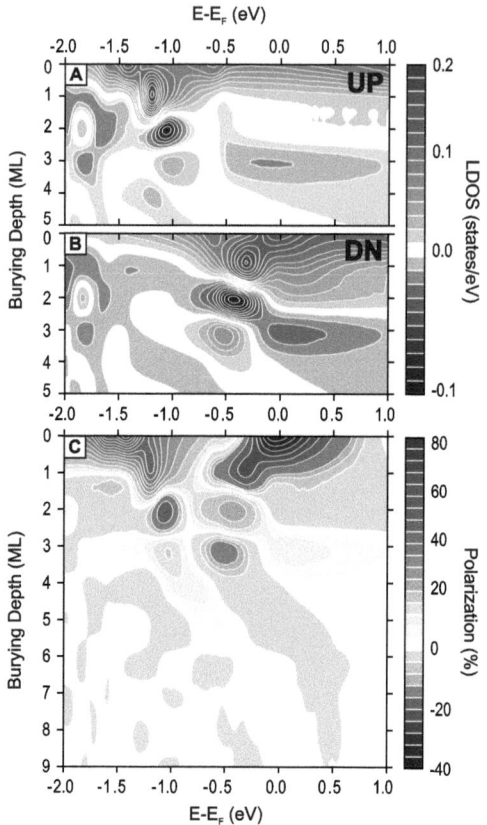

Figure 5.4: Energy resolved majority (A) and minority (B) LDOS (color-coded) in vacuum above the embedding site of a single Co adatom versus the burying depth. The LDOS of a host surface has been subtracted for clarity. (C) Polarization (color-coded) $P = P(E, d)$ above a single buried Co adatom as a function of electron energy and burying depth. The $P(E)$ distributions have been calculated for integer layer numbers and then interpolated for clarity.

5.1 Probing the electronic and magnetic properties of buried nanostructures

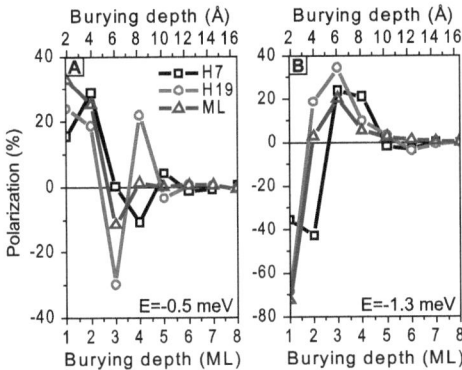

Figure 5.5: Comparison of the polarization at the surface above buried 7-atomic (H7,black squares), 19-atomic (H19, red circles) clusters and a buried monolayer (ML, blue triangles pointing up) at −0.5 eV (A) and −1.3 eV (B). The lines are meant solely as a guide for the eye. From

larger than the lateral extents of both clusters, the corresponding polarization curves display a very similar behavior. For shallow-buried clusters, when the size of the structure comes into play, the behavior of the curves begins to differ considerably. Another asymptotic feature, that can be noticed is the striking qualitative similarity between the H19 and ML curves in the depth range of 6 − 7 Å. It means that at shallow burying depths the H19 cluster influences the surface polarization above it's burying site in a way very similar to that of a complete monolayer. This might be of use when contemplating possible experimental or technological applications, as it marks the lateral extents for a single magnetic unit which would provide us with an imitation of a monolayer. Magnetic monolayers (or, to be precise, stacks of them) are currently the basis for many magnetic devices relying on giant and tunneling magnetoresistance and similar techniques.

One more issue, that is bound to arise in an experimentalist's mind, is the question of the stability of the cluster's magnetic orientation. It is well known (see f.e. [195]) that small metallic clusters at surfaces often exhibit superparamagnetic properties which would render observations discussed above virtually impossible. However, as remarked in , the fact that at large burying depths the role of the cluster's size ceases to play a major role, could make it possible, by increasing the size of the clusters, to achieve sufficient values of magnetic anisotropy to sustain a constant direction of the magnetic moment at reasonable experimental environmental conditions. Alternatively a small magnetic field might serve as a stabilizing factor. It would, however, also interfere with the intrinsic magnetic interactions in the system.

Although by looking at a single point above a buried magnetic nanostructure one can already obtain some information about the structure's burying depth and shed some light onto its electronic and magnetic properties, it is obvious that a space-resolved scan of the polarization distribution in vacuum above the surface would give one a much deeper insight. An example of such polarization scans calculated at −0.5 eV in vacuum above an H7 cluster buried in monolayers 1 to 4 (A to D respectively) beneath the surface are presented as a function of surface in-plane coordinates in Fig. 5.6. A buried cluster leaves a unique

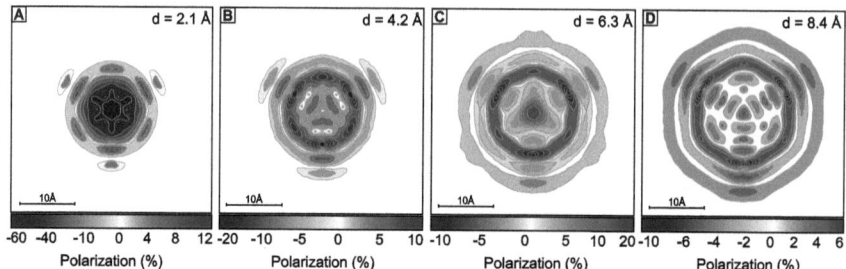

Figure 5.6: Polarization maps above a hexagonal 7-atomic cluster of Co residing under a Cu(111) surface at burying depths of 2.1 (A), 4.2 (B), 6.3 (C) and 8.4 (D) Å.

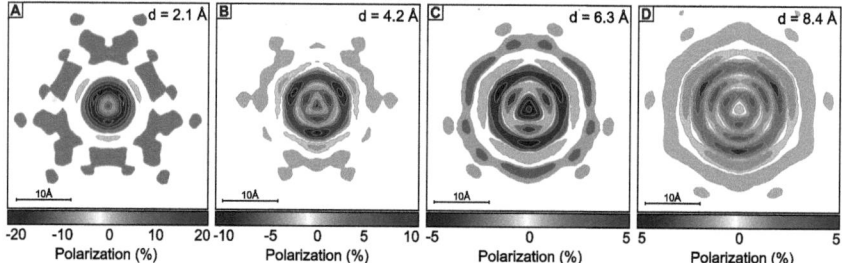

Figure 5.7: Polarization maps above a single Co adatom residing under a Cu(111) surface at burying depths of 2.1 (A), 4.2 (B), 6.3 (C) and 8.4 (D) Å.

polarization imprint in the LDOS, and hence the polarization, in the vacuum space above it's burying site for each burying depth. A critical look at the figure immediately reveals a 3-fold rotational symmetry in the polarization maps, the origin of which, however, is easily understood if one considers the intrinsic geometry of a (111) surface. Looking at the polarization distributions one once again detects the characteristic details of the electronic interference, namely the radial oscillations of the polarization. Their origin is also quite easily understood: the phase relation of the incoming and scattered electronic waves change, in accordance with simple geometric laws, causing the resulting periodic variations in the polarization of surface electrons. Note also that with increasing burying depth the radial period of the oscillations decreases which also complies with simple notions of geometric optics. The phase of the oscillations is determined by the electronic properties of impurity and host materials as well as by the burying depth of the impurity. Consequently, the phase and the period of in-plane radial oscillations of the polarization can provide us with important information about the position and the burying depth of an embedded nanostructure with known electronic properties.

Similar distributions for a simpler system, a single Co adatom buried in monolayers 1 to 4 (A to D respectively) beneath the Cu(111) surface are presented in Fig. 5.7. The same trends as exist in Fig. 5.6 are also clearly traceable here.

Of course, should this topic would motivate a further experimental study, the theoretical investigation can be extended to include the calculation of experimental geometries and the

5.1 Probing the electronic and magnetic properties of buried nanostructures   89

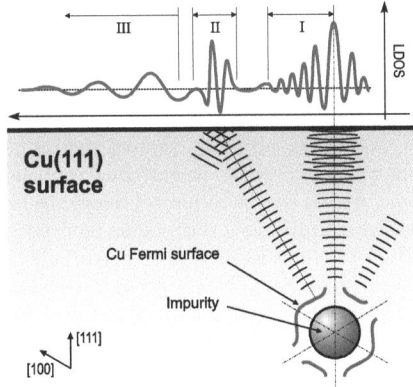

Figure 5.8: A sketch of possible sources of the surface LDOS perturbation. A Cu(111) surface with an embedded Co impurity. Concentric rings depict the preferable directions of electron propagations. The curve above the surface presents an abstract sketch of a LDOS distribution along a single direction of the surface. Region (I) is the area just above the impurity, where the LDOS is influenced by the quasi spherical electron waves emitted from the burying site. Region (II) marks the area where the quasi plane waves emitted from the impurity's burying site in the directions determined by the regions of strongly reduced Fermi-surface-curvature reach the surface. Region (III) is a generalization of the areas, where the Friedel oscillations in the Cu(111) surface state density caused by the presence of the buried impurity exceed in amplitude all the other contributions.

results should be formulated (for easier comparison) in terms of spin-resolved differential conductance (based, f.e., on the Tersoff-Hamman model [49]).

For completeness sake, it must also be noted here, that for a possible comparison with an experiment care should be taken to select a suitable energy window, as the LDOS perturbation at the surface is very much energy dependent. To clarify the point, let us consider possible ways in which a buried impurity might influence the LDOS, and hence the polarization, of the surface. In the first approximation three different radial regions (with respect to the projection of the burying site onto the surface) might be defined (Fig. 5.8). The first one (I) is located just above the impurity, where the LDOS is influenced by the quasi spherical electron waves emitted from the burying site. These are the bulk electrons scattered by the impurity orbitals in the direction vicinal to the [111] vector. The changes in the LDOS at the surface are completely determined by the Green's function of the system [190]

$$\Delta \text{LDOS}(x, E) = -\frac{1}{\pi}\Im \iint G(x, x_i, E) t(x_i, x_j, E) G(x_j, x, E) dx_i dx_j,$$

where x_i, x_j are arbitrary coordinates within the system and t is the t-matrix of the impurity. Thus in region (I) one might expect the perturbative contribution from the quasi free electrons propagating in and around (in the reciprocal sense) the bulk band gap. The radial decay of the Green's function with distance then might be expected to be of an exponential character (see Subsection 1.7.3 for details). Such perturbation would be thus decaying extremely fast with increasing burying depth. According to considerations presented in Section 1.4 and Ref. [190] the decay of the electronic waves propagating along a certain direction would be inversely proportional to the Gaussian curvature of the corresponding patch of the isoenergy surface (for energy ε). For the patches of the isoenergy surface with a highly reduced curvature, this perturbations would propagate virtually without decay and can be sensed even at large distances from the impurity. They are, thus, expected to strongly perturb the LDOS at the surface at corresponding energies [190]. The region of the surface where such quasi plane wave perturbations intersect with the surface plane are marked as (II) in Fig. 5.8. Region (III) is a generalization of those areas, where the Friedel oscillations in the Cu(111) surface state density caused by the presence of the buried impurity (see, f.e. [183, 181, 186]) exceed in amplitude all the other contributions. Such oscillations are also expected to carry a spin-polarized character for magnetic subsurface impurities. Though the three regions might be theoretically distinct, it is virtually impossible to clearly separate them in a real experiment or calculation: at shallow burying depths regions I and II will overlap creating a complex perturbation distribution (or polarization distribution) similar to Fig. 5.6 and 5.7. At greater burying depths region I would be nigh impossible to detect due to the exponential decay of the corresponding Green's function, and the Friedel oscillations will provide a smooth background for region II, as the Fermi wave length of surface state electrons is much larger than that of their bulk counterparts. To detect the long-range Friedel oscillations one will have to map the surface with an STM on a much larger scale.

We thus believe, that magnetic properties of nanostructures buried beneath a metallic surface can be deduced from the spin-resolved local density of states above the surface. Acquiring in-plane polarization maps in vacuum above the surface can allow one to simultaneously detect electronic, magnetic and even geometric properties of subsurface structures. The possibility of deducing the geometric structure of a subsurface impurity is however still in need of further investigation, e.g. by fitting the polarization distribution maps at the surface with a simple multiple scattering model.

5.2 Checking the coupling of buried nanostructures

An important step towards real applications of subsurface impurities is gaining the understanding of the nature of their interaction. To make this step one obviously needs a way to effectively probe the interaction between single buried nanostructures. Let us consider a pair of H7 clusters of Co buried at the same monolayer 6.3 Å deep beneath a Cu(111) surface with a center-center separations of 20 Å. The strong ferromagnetic nature of Co would definitely align all the spins inside each of the Co clusters parallel to each other. So it is the relative alignment of the two clusters' spins which is *a priori* unknown and depends on many factors, such as the islands' relative position and the host material of the surface. However, if one were to probe the polarization distribution on the surface above the two clusters one might obtain a picture similar to that presented in Fig. 5.9(FM and AFM) for the case of parallel (FM) and antiparallel (AFM) orientation of clusters' magnetic moments respectively. Figures presented, cover a separation range of 7 − 20 Å. For all separations, the system with a FM alignment of moments produces a symmetric polarization map which is a superposition of two similar distributions set off from each other by a distance equal to the cluster separation d. The AFM system, on the contrary, would produce a map which is perfectly antisymmetrical with respect to the symmetry plane $(\bar{1}01)$ separating the buried structures. Such a behavior can easily be understood if one considers that in a system with AFM orientation of moments the electrons scattered at majority states of one cluster will interfere with electrons scattered at minority states of the other, thus creating an antisymmetrical LDOS distribution at the surface. So we arrive at the conclusion that the symmetry of the polarization map can be regarded as a signature of the relative moment orientation of buried clusters. Moreover, it can be noted that the antisymmetrical polarization distribution implies the absence of polarization along the symmetry plane (or above the symmetry center of the map marked by crosses in Fig. 5.9) which might be regarded as an additional and even simpler criterion of the clusters' moments alignment.

5.3 Outlook

We have thus achieved what we set aut to achieve: we have shown, that the coupling of buried nanostructures to each other can be deduced from the symmetry of the polarization map. Such measurements can be carried out by means of a conventional spin-polarized STM. It is also important to note that even if the resolution of a modern STM should prove insufficient for high resolution polarization mapping, one might still study the magnetization of the surface utilizing the notions from Section 3.1.

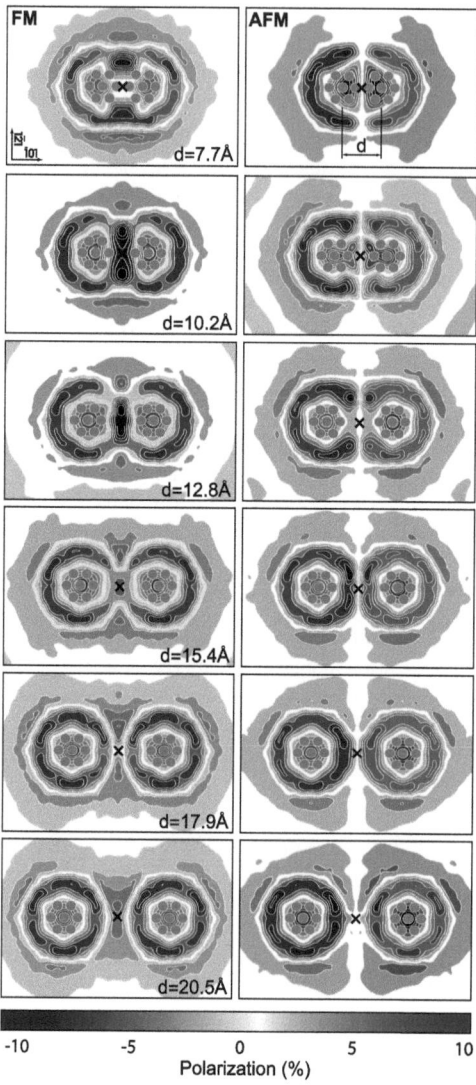

Figure 5.9: Polarization above a pair of H7 clusters of Co buried in the same layer 6.3 Å deep beneath a Cu(111) surface with a center-center separations ranging between 7.7 and 20.5 Å and either a parallel (FM) or antiparallel (AFM) alignment of clusters' magnetic moments. Red and blue circles denote the burying sites of Co atoms with spin pointing up and down respectively. All the maps are plotted for the electrons at $E = -0.5$ eV.

Conclusions

To outline once again the presented results we might conclude, that our *ab initio* calculations, carried out with the Korringa-Kohn-Rostoker Green's function method, give us confidence that it is possible to tune and probe magnetic interaction in sub-nanoscale structures at metallic surfaces.

Coupling single magnetic structures or structure ensembles at the surface to buried magnetic layers one obtains (by deliberately changing the layer's burying depth) unique control over the coupling of the surface structures to the monolayer and to each other. Using this technique, the spins of single atoms can be stabilized in either a ferro- or antiferromagnetic configuration with respect to the magnetic moment of the monolayer and the coupling of such atoms under each other can be switched. It is also possible to additionally manipulate the exchange interaction between magnetic adatoms at a surface by introducing artificial nonmagnetic chains to link them.

The influence of quantum confinement of surface electrons on the exchange interaction between single adatoms has been shown to enable exchange interaction tuning in pairs of adatoms adsorbed on hexagonal nanoscale paramagnetic islands: a system easily obtained by selfassembly at surfaces of some of the transitional metals. It was demonstrated that it is possible to enhance, reduce or even reverse the exchange coupling at various adatom-adatom separations by deliberate choice of the island's size.

The dependence of width and energetic position of spin-polarized bound states arising at pairs of magnetic adatoms on the spin coupling in the system has been discussed. The possibility to use the spin splitting of the bound state peak as a tool for probing the exchange coupling in the system was pointed out.

Finally, we have shown that it is possible to resolve magnetic properties of small clusters buried beneath a metallic surface by studying the polarization of electrons in vacuum space above the system. Moreover, the same technique could allow one to determine the coupling of buried structures to each other.

Bibliography

[1] G. Binasch, P. Grünberg, F. Saurenbach and W. Zinn. *Enhanced magnetoresistance in layered magnetic structures with antiferromagnetic interlayer exchange.* Phys. Rev. B, **39**, 4828 (1989).

[2] M. N. Baibich, J. M. Broto, A. Fert, F. N. Van Dau, F. Petroff, P. Eitenne, G. Creuzet, A. Friederich and J. Chazelas. *Giant Magnetoresistance of (001)Fe/(001)Cr Magnetic Superlattices.* Phys. Rev. Lett., **61**, 2472 (1988).

[3] S. A. Wolf, D. D. Awschalom, R. A. Buhrman, J. M. Daughton, S. von Molnar, M. L. Roukes, A. Y. Chtchelkanova and D. M. Treger. *Spintronics: A Spin-Based Electronics Vision for the Future.* Science, **294**, 1488 (2001).

[4] F. H. L. Koppens, J. A. Folk, J. M. Elzerman, R. Hanson, L. H. W. van Beveren, I. T. Vink, H. P. Tranitz, W. Wegscheider, L. P. Kouwenhoven and L. M. K. Vandersypen. *Control and Detection of Singlet-Triplet Mixing in a Random Nuclear Field.* Science, **309**, 1346 (2005).

[5] J. R. Petta, A. C. Johnson, J. M. Taylor, E. A. Laird, A. Yacoby, M. D. Lukin, C. M. Marcus, M. P. Hanson and A. C. Gossard. *Coherent Manipulation of Coupled Electron Spins in Semiconductor Quantum Dots.* Science, **309**, 2180 (2005).

[6] V. S. Stepanyuk, A. N. Baranov, W. Hergert and P. Bruno. *Ab initio study of interaction between magnetic adatoms on metal surfaces.* Phys. Rev. B, **68**, 205422 (2003).

[7] X. Xu, S. Yin, R. Moro and W. A. de Heer. *Magnetic Moments and Adiabatic Magnetization of Free Cobalt Clusters.* Phys. Rev. Lett., **95**, 237209 (2005).

[8] P. Gambardella, A. Dallmeyer, K. Maiti, M. Malagoli, W. Eberhardt, K. Kern and C. Carbone. *Ferromagnetism in one-dimensional monatomic metal chains.* Nature, **416**, 301 (2002).

[9] K. Tao, V. S. Stepanyuk, W. Hergert, I. Rungger, S. Sanvito and P. Bruno. *Switching a Single Spin on Metal Surfaces by a STM Tip: Ab Initio Studies.* Phys. Rev. Lett., **103**, 057202 (2009).

[10] J. Barnaś and Y. Bruynseraede. *Correlation between quantum-size effects in the giant magnetoresistance and interlayer coupling in magnetic multilayers.* Phys. Rev. B, **53**, R2956 (1996).

[11] P. Bruno and C. Chappert. *Oscillatory coupling between ferromagnetic layers separated by a nonmagnetic metal spacer.* Phys. Rev. Lett., **67**, 1602 (1991).

[12] P. Bruno and C. Chappert. *Ruderman-Kittel theory of oscillatory interlayer exchange coupling*. Phys. Rev. B, **46**, 261 (1992).

[13] H. A. Mizes and J. S. Foster. *Long-Range Electronic Perturbations Caused by Defects Using Scanning Tunneling Microscopy*. Science, **244**, 559 (1989).

[14] M. Schmid, W. Hebenstreit, P. Varga and S. Crampin. *Quantum Wells and Electron Interference Phenomena in Al due to Subsurface Noble Gas Bubbles*. Phys. Rev. Lett., **76**, 2298 (1996).

[15] H. Brune, J. Wintterlin, G. Ertl and R. J. Behm. *Direct Imaging of Adsorption Sites and Local Electronic-Bond Effects on a Metal-Surface - C/AL(111)*. Europhys. Lett., **13**, 123 (1990).

[16] M. F. Crommie, C. P. Lutz and D. M. Eigler. *Confinement of Electrons to Quantum Corrals on a Metal Surface*. Science, **262**, 218 (1993).

[17] L. C. Davis, M. P. Everson, R. C. Jaklevic and W. D. Shen. *Theory of The Local Density of Surface-States on a Metal - Comparison With Scanning Tunneling Spectroscopy of a Au(111) Surface*. Phys. Rev. B, **43**, 3821 (1991).

[18] M. F. Crommie, C. P. Lutz and D. M. Eigler. *Imaging Standing Waves in a 2-Dimensional Electron-Gas*. Nature, **363**, 524 (1993).

[19] Y. Hasegawa and P. Avouris. *Direct observation of standing wave formation at surface steps using scanning tunneling spectroscopy*. Phys. Rev. Lett., **71**, 1071 (1993).

[20] L. Petersen, P. Laitenberger, E. Laegsgaard and F. Besenbacher. *Screening waves from steps and defects on Cu(111) and Au(111) imaged with STM: Contribution from bulk electrons*. Phys. Rev. B, **58**, 7361 (1998).

[21] J. Koutecky. *A Contribution to the Molecular-Orbital Theory of Chemisorption*. Transactions of the Faraday Society, **54**, 1038 (1958).

[22] K. H. Lau and W. Kohn. *Indirect long-range oscillatory interaction between adsorbed atoms*. Surf. Sci., **75**, 69 (1978).

[23] J. Repp, F. Moresco, G. Meyer, K.-H. Rieder, P. Hyldgaard and M. Persson. *Substrate Mediated Long-Range Oscillatory Interaction between Adatoms: Cu /Cu(111)*. Phys. Rev. Lett., **85**, 2981 (2000).

[24] N. Knorr, H. Brune, M. Epple, A. Hirstein, M. A. Schneider and K. Kern. *Long-range adsorbate interactions mediated by a two-dimensional electron gas*. Phys. Rev. B, **65**, 115420 (2002).

[25] L. Bürgi, N. Knorr, H. Brune, M. A. Schneider and K. Kern. *Two-dimensional electron gas at noble-metal surfaces*. Appl. Phys. A: Mater. Sci. Process., **75**, 141 (2002).

[26] V. S. Stepanyuk, A. N. Baranov, D. V. Tsivlin, W. Hergert, P. Bruno, N. Knorr, M. A. Schneider and K. Kern. *Quantum interference and long-range adsorbate-adsorbate interactions*. Phys. Rev. B, **68**, 5 (2003).

[27] O. Braun and V. Medvedev. *Interaction between particles adsorbed on metal surfaces.* Uspekhi Fiz. Nauk, **157**, 631 (1989).

[28] O. M. Braun and V. K. Medvedev. *Interaction between particles adsorbed on metal surfaces.* Sov. Phys. Usp., **32**, 328 (1989).

[29] E. J. Heller, M. F. Crommie, C. P. Lutz and D. M. Eigler. *Scattering and absorption of surface electron waves in quantum corrals.* Nature, **369**, 464 (1994).

[30] H. C. Manoharan, C. P. Lutz and D. M. Eigler. *Quantum mirages formed by coherent projection of electronic structure.* Nature, **403**, 512 (2000).

[31] J. Li, W.-D. Schneider, R. Berndt and B. Delley. *Kondo Scattering Observed at a Single Magnetic Impurity.* Phys. Rev. Lett., **80**, 2893 (1998).

[32] V. Madhavan, W. Chen, T. Jamneala, M. Crommie and N. Wingreen. *Tunneling into a single magnetic atom: Spectroscopic evidence of the Kondo resonance.* Science, **280**, 567 (1998).

[33] V. S. Stepanyuk, L. Niebergall, W. Hergert and P. Bruno. *Ab initio Study of Mirages and Magnetic Interactions in Quantum Corrals.* Phys. Rev. Lett., **94**, 187201 (2005).

[34] V. S. Stepanyuk, N. N. Negulyaev, L. Niebergall and P. Bruno. *Effect of quantum confinement of surface electrons on adatom-adatom interactions.* New J. Phys., **9**, 15 (2007).

[35] N. N. Negulyaev, V. S. Stepanyuk, L. Niebergall, P. Bruno, W. Hergert, J. Repp, K.-H. Rieder and G. Meyer. *Direct Evidence for the Effect of Quantum Confinement of Surface-State Electrons on Atomic Diffusion.* Phys. Rev. Lett., **101**, 226601 (2008).

[36] A. A. Correa, F. A. Reboredo and C. A. Balseiro. *Quantum corral wave-function engineering.* Phys. Rev. B, **71**, 035418 (2005).

[37] C. Trallero-Giner, S. E. Ulloa and V. López-Richard. *Local density of states in parabolic quantum corrals.* Phys. Rev. B, **69**, 115423 (2004).

[38] J. R. Friedman, M. P. Sarachik, J. Tejada and R. Ziolo. *Macroscopic Measurement of Resonant Magnetization Tunneling in High-Spin Molecules.* Phys. Rev. Lett., **76**, 3830 (1996).

[39] L. Thomas, F. Lionti, R. Ballou, D. Gatteschi, R. Sessoli and B. Barbara. *Macroscopic quantum tunnelling of magnetization in a single crystal of nanomagnets.* Nature, **383**, 145 (1996).

[40] S. Hill, R. Edwards, N. Aliaga-Alcalde and G. Christou. *Quantum coherence in an exchange-coupled dimer of single-molecule magnets.* Science, **302**, 1015 (2003).

[41] R. Caciuffo, G. Amoretti, A. Murani, R. Sessoli, A. Caneschi and D. Gatteschi. *Neutron Spectroscopy for the Magnetic Anisotropy of Molecular Clusters.* Phys. Rev. Lett., **81**, 4744 (1998).

[42] G. Binnig, H. Rohrer, C. Gerber and E. Weibel. *Surface Studies by Scanning Tunneling Microscopy.* Phys. Rev. Lett., **49**, 57 (1982).

[43] G. Binnig and H. Rohrer. *Scanning tunneling microscopy - from birth to adolescence.* Rev. Mod. Phys., **59**, 615 (1987).

[44] G. Binnig, H. Rohrer, C. Gerber and E. Weibel. *7×7 Reconstruction on Si(111) Resolved in Real Space.* Phys. Rev. Lett., **50**, 120 (1983).

[45] R. Wiesendanger, H.-J. Güntherodt, G. Güntherodt, R. J. Gambino and R. Ruf. *Observation of vacuum tunneling of spin-polarized electrons with the scanning tunneling microscope.* Phys. Rev. Lett., **65**, 247 (1990).

[46] M. Bode, M. Getzlaff and R. Wiesendanger. *Spin-Polarized Vacuum Tunneling into the Exchange-Split Surface State of Gd(0001).* Phys. Rev. Lett., **81**, 4256 (1998).

[47] R. Wiesendanger. *Scanning Probe Microscopy, Chap. 4.* Springer, Berlin (1998).

[48] J. Tersoff and D. R. Hamann. *Theory and Application for the Scanning Tunneling Microscope.* Phys. Rev. Lett., **50**, 1998 (1983).

[49] J. Tersoff and D. R. Hamann. *Theory of the scanning tunneling microscope.* Phys. Rev. B, **31**, 805 (1985).

[50] J. Bardeen. *Tunnelling from a Many-Particle Point of View.* Phys. Rev. Lett., **6**, 57 (1961).

[51] D. Wortmann, S. Heinze, P. Kurz, G. Bihlmayer and S. Blügel. *Resolving Complex Atomic-Scale Spin Structures by Spin-Polarized Scanning Tunneling Microscopy.* Phys. Rev. Lett., **86**, 4132 (2001).

[52] R. Wiesendanger. *Spin mapping at the nanoscale and atomic scale.* Rev. Modern Phys., **81**, 1495 (2009).

[53] W. Wulfhekel and J. Kirschner. *Spin-polarized scanning tunneling microscopy on ferromagnets.* Appl. Phys. Lett., **75**, 1944 (1999).

[54] C. F. Hirjibehedin, C. P. Lutz and A. J. Heinrich. *Spin Coupling in Engineered Atomic Structures.* Science, **312**, 1021 (2006).

[55] D. Eigler and E. Schweizer. *Positioning Single Atoms With a Scanning Tunneling Microscope.* Nature, **344**, 524 (1990).

[56] A. Heinrich, J. Gupta, C. Lutz and D. Eigler. *Single-atom spin-flip spectroscopy.* Science, **306**, 466 (2004).

[57] W. Chen, T. Jamneala, V. Madhavan and M. F. Crommie. *Disappearance of the Kondo resonance for atomically fabricated cobalt dimers.* Phys. Rev. B, **60**, R8529 (1999).

[58] P. Wahl, P. Simon, L. Diekhöner, V. S. Stepanyuk, P. Bruno, M. A. Schneider and K. Kern. *Exchange Interaction between Single Magnetic Adatoms.* Phys. Rev. Lett., **98**, 056601 (2007).

[59] A. C. Hewson. *The Kondo Problem to Heavy Fermions.* Cambridge University Press, Cambridge, England (1993).

[60] P. Wahl, L. Diekhoner, G. Wittich, L. Vitali, M. Schneider and K. Kern. *Kondo effect of molecular complexes at surfaces: Ligand control of the local spin coupling.* Phys. Rev. Lett., **95** (2005).

[61] U. Fano. *Effects of configuration interaction on intensities and phase shifts.* Phys. Rev., **124**, 1866 (1961).

[62] P. Wahl, L. Diekhoner, M. Schneider, L. Vitali, G. Wittich and K. Kern. *Kondo temperature of magnetic impurities at surfaces.* Phys. Rev. Lett., **93** (2004).

[63] V. S. Stepanyuk, A. N. Baranov, D. I. Bazhanov, W. Hergert and A. A. Katsnelson. *Magnetic properties of mixed Co-Cu clusters on Cu(0 0 1).* Surf. Sci., **482-485**, 1045 (2001).

[64] C. Jayaprakash, H. R. Krishna-murthy and J. W. Wilkins. *Two-Impurity Kondo Problem.* Phys. Rev. Lett., **47**, 737 (1981).

[65] R. López, R. Aguado and G. Platero. *Nonequilibrium Transport through Double Quantum Dots: Kondo Effect versus Antiferromagnetic Coupling.* Phys. Rev. Lett., **89**, 136802 (2002).

[66] P. Simon, R. López and Y. Oreg. *Ruderman-Kittel-Kasuya-Yosida and Magnetic-Field Interactions in Coupled Kondo Quantum Dots.* Phys. Rev. Lett., **94**, 086602 (2005).

[67] F. Meier, L. Zhou, J. Wiebe and R. Wiesendanger. *Revealing Magnetic Interactions from Single-Atom Magnetization Curves.* Science, **320**, 82 (2008).

[68] P. Gambardella, S. Rusponi, M. Veronese, S. Dhesi, C. Grazioli, A. Dallmeyer, I. Cabria, R. Zeller, P. Dederichs, K. Kern, C. Carbone and H. Brune. *Giant magnetic anisotropy of single cobalt atoms and nanoparticles.* Science, **300**, 1130 (2003).

[69] M. A. Ruderman and C. Kittel. *Indirect Exchange Coupling of Nuclear Magnetic Moments by Conduction Electrons.* Phys. Rev., **96**, 99 (1954).

[70] J. Friedel. *Electronic structure of primary solid solutions in metals.* Adv. Phys., **50**, 539 (2001).

[71] J. Friedel. *Electronic structure of primary solid solutions in metals.* Adv. in Phys., **3**, 446 (1954).

[72] J. Langer and S. Vosko. *The shielding of a fixed charge in a high-density electron gas.* J. Phys. and Chem. of Sol., **12**, 196 (1960).

[73] E. Canel, M. P. Matthews and R. K. P. Zia. *Screening in very thin-films.* Phys. der Kond. Mat., **15**, 191 (1972).

[74] A. K. Das. *The Linear Response of a One-Dimensional Electron Gas.* Sol. St. Commun., **15**, 475 (1974).

[75] T. B. Grimley. *The indirect interaction between atoms or molecules adsorbed on metals.* Proc. Phys. Soc., **90**, 751 (1967).

[76] T. B. Grimley. *The electron density in a metal near a chemisorbed atom or molecule.* Proc. Phys. Soc., **92**, 776 (1967).

[77] T. Grimley and S. Walker. *Interactions between adatoms on metals and their effects on the heat of adsorption at low surface coverage.* Surf. Sci., **14**, 395 (1969).

[78] A. M. Gabovich. Fiz. Tverd. Tela, **18**, 377 (1976).

[79] A. M. Gabovich. Sov. Phys. Sol. St., **18**, 220 (1976).

[80] D. M. Newns. *Self-Consistent Model of Hydrogen Chemisorption.* Phys. Rev., **178**, 1123 (1969).

[81] P. W. Anderson. *Localized Magnetic States in Metals.* Phys. Rev., **124**, 41 (1961).

[82] R. W. Gurney. *Theory of Electrical Double Layers in Adsorbed Films.* Phys. Rev., **47**, 479 (1935).

[83] O. M. Braun. Ukrainskiĭ fizicheskiĭ zhurnal, **23**, 1233 (1978).

[84] T. L. Einstein and J. R. Schrieffer. *Indirect Interaction between Adatoms on a Tight-Binding Solid.* Phys. Rev. B, **7**, 3629 (1973).

[85] L. D. Landau, E. M. Lifshits and L. P. Pitaevskii. *Statystical Physics, Part 2.* Butterworth-Heinemann (1980).

[86] A. Abrikosov, L. Gorkov and I. Dzyaloshinski. *Methods of Quantum Field Theory in Statistical Physics.* Englewood Cliffs (1963).

[87] A. Abrikosov, L. Gorkov and I. Dzyaloshinski. *Methods of Quantum Field Theory in Statistical Physics, 3rd Edition.* KDU, Dobrosvet (2006).

[88] T. L. Einstein. *Theory of Indirect Interaction Between Chemisorbed Atoms.* Crc Crit. Rev. Sol. St. and Mat. Sci., **7**, 261 (1978).

[89] J. W. Gadzuk. *Surface Physics Materials, vol. 2.* Academic Press, New York and London, 1975 (1975).

[90] J. P. Muscat and D. M. Newns. *Chemisorption on metals.* Progress in Surf. Sci., **9**, 1 (1978).

[91] E. M. Lifshits and L. P. Pitaevskii. *Statisticheskaya Fizika, Part 2: Solid state physics.* Fizmatlit, Moscow (2004).

[92] P. Johansson. *Electron density oscillations around an adatom.* Sol. St. Commun., **31**, 591 (1979).

[93] T. L. Einstein. *Comment on oscillatory indirect interaction between adsorbed atoms – Non-asymptotic behavior in tight-binding models at realistic parameters by K.H. Lau and W. Kohn.* Surf. Sci., **75**, 161 (1978).

[94] O. Braun. *Some Peculiarities of Indirect Interaction of Atoms Absorbed on Metal-Surface.* Fiz. Tverd. Tela, **23**, 2779 (1981).

[95] J. L. Bosse, J. Lopez and J. Rousseau-Violet. *Interaction energy between two identical atoms chemisorbed on a normal metal.* Surf. Sci., **72**, 125 (1978).

[96] A. Brodskii and M. Urbakh. *On the interactions of adatoms on metal surfaces.* Surf. Sci., **105**, 196 (1981).

[97] I. Tamm. *Über eine mögliche Art der Elektronenbindung an Kristalloberflächen.* Zeitschrift fur Physik, **76**, 849 (1932).

[98] W. Shockley. *On the Surface States Associated with a Periodic Potential.* Phys. Rev., **56**, 317 (1939).

[99] E. Wahlström, I. Ekvall, H. Olin and L. Walldén. *Long-range interaction between adatoms at the Cu(111) surface imaged by scanning tunnelling microscopy.* Appl. Phys. A: Mater. Sci. Process., **66**, 1107 (1998).

[100] P. Hyldgaard and M. Persson. *Long-ranged adsorbate-adsorbate interactions mediated by a surface-state band.* J. Phys. Cond. Matt., **12**, L13 (2000).

[101] A. Bogicevic, S. Ovesson, P. Hyldgaard, B. I. Lundqvist, H. Brune and D. R. Jennison. *Nature, Strength, and Consequences of Indirect Adsorbate Interactions on Metals.* Phys. Rev. Lett., **85**, 1910 (2000).

[102] P. Hyldgaard and T. L. Einstein. *Surface-state–mediated three-adsorbate interaction.* Europhys. Lett., **59**, 265 (2002).

[103] F. Silly, M. Pivetta, M. Ternes, F. Patthey, J. P. Pelz and W.-D. Schneider. *Creation of an Atomic Superlattice by Immersing Metallic Adatoms in a Two-Dimensional Electron Sea.* Phys. Rev. Lett., **92**, 016101 (2004).

[104] H. F. Ding, V. S. Stepanyuk, P. A. Ignatiev, N. N. Negulyaev, L. Niebergall, M. Wasniowska, C. L. Gao, P. Bruno and J. Kirschner. *Self-organized long-period adatom strings on stepped metal surfaces: Scanning tunneling microscopy, ab initio calculations, and kinetic Monte Carlo simulations.* Phys. Rev. B, **76**, 033409 (2007).

[105] A. S. Smirnov, N. N. Negulyaev, W. Hergert, A. M. Saletsky and V. S. Stepanyuk. *Magnetic behavior of one- and two-dimensional nanostructures stabilized by surface-state electrons: a kinetic Monte Carlo study.* New J. Phys., **11**, 063004 (16pp) (2009).

[106] N. N. Negulyaev, V. S. Stepanyuk, L. Niebergall, P. Bruno, M. Pivetta, M. Ternes, F. Patthey and W.-D. Schneider. *Melting of Two-Dimensional Adatom Superlattices Stabilized by Long-Range Electronic Interactions.* Phys. Rev. Lett., **102**, 246102 (2009).

[107] S. Alexander and P. W. Anderson. *Interaction Between Localized States in Metals.* Phys. Rev., **133**, A1594 (1964).

[108] A. Oswald, R. Zeller, P. J. Braspenning and P. H. Dederichs. *Interaction of magnetic impurities in Cu and Ag.* J. Phys. F: Met. Phys., **15**, 193 (1985).

[109] T. Kasuya. *Electrical Resistance of Ferromagnetic Metals.* Prog. Theor. Phys., **16**, 58 (1956).

[110] K. Yosida. *Magnetic Properties of Cu-Mn Alloys.* Phys. Rev., **106**, 893 (1957).

[111] J. Friedel. *Metallic alloys.* Nuovo Cimento, **7**, 287 (1958).

[112] P. W. Anderson. *New Approach to the Theory of Superexchange Interactions.* Phys. Rev., **115**, 2 (1959).

[113] J. B. Staunton, B. L. Gyorffy, J. Poulter and P. Strange. *A relativistic RKKY interaction between two magnetic impurities-the origin of a magnetic anisotropic effect.* J. Phys. C: Sol. St. Phys., **21**, 1595 (1988).

[114] A. Fert and P. M. Levy. *Role of Anisotropic Exchange Interactions in Determining the Properties of Spin-Glasses.* Phys. Rev. Lett., **44**, 1538 (1980).

[115] S. M. Goldberg, P. M. Levy and A. Fert. *Anisotropy in binary metallic spin-glass alloys. I. Transition metals.* Phys. Rev. B, **33**, 276 (1986).

[116] I. Dzyaloshinsky. *A thermodynamic theory of weak ferromagnetism of antiferromagnetics.* J. Phys. and Chem. of Sol., **4**, 241 (1958).

[117] T. Moriya. *New Mechanism of Anisotropic Superexchange Interaction.* Phys. Rev. Lett., **4**, 228 (1960).

[118] T. Moriya. *Anisotropic Superexchange Interaction and Weak Ferromagnetism.* Phys. Rev., **120**, 91 (1960).

[119] P. Bruno. *Theory of interlayer exchange interactions in magnetic multilayers.* J. Phys.: Cond. Matt., **11**, 9403 (1999).

[120] P. Bruno. J. Magn. Magn. Mat., **164**, 27 (1996).

[121] P. Bruno. *Theory of interlayer magnetic coupling.* Phys. Rev. B, **52**, 411 (1995).

[122] P. Bruno. *Oscillations of Interlayer Exchange Coupling vs. Ferromagnetic-Layers Thickness.* Europhys. Lett., **23**, 615 (1993).

[123] P. Bruno. J. Magn. Magn. Mat., **121**, 248 (1993).

[124] A. Katsnelson, V. Stepanyuk, A. Szasz and O. Farberovich. *Computational methods in condensed matter; electronic structure.* AIP, New York, NY, USA (1992).

[125] R. M. Martin. *Electronic structure: basic theory and practical methodsElectronic structure: basic theory and practical methods.* Cambridge university press (2004).

[126] P. Hohenberg and W. Kohn. *Inhomogeneous electron gas.* Phys. Rev., **136**, B864 (1964).

[127] E. K. Gross and R. M. Dreizler. *Density Functional Theory*, vol. Vol. 337. NATO Science Series B: Physics (1995).

[128] D. Joubert. *Density Functionals: Theory and Applications*, vol. Vol. 500 of *Series: Lecture Notes in Physics.* Springer (1998). ISBN: 978-3-540-63937-4.

[129] P. Weinberger. *Electron scattering theory for ordered and disordered matter*. Clarendon Press, Oxford University Press (1990).

[130] W. Kohn and L. J. Sham. Phys. Rev., **140**, A1133 (1965).

[131] J. Korringa. *On the calculation of the energy of a bloch wave in a metal*. Physica, **13**, 392 (1947).

[132] W. Kohn and N. Rostoker. *Solution of the Schrödinger Equation in Periodic Lattices with an Application to Metallic Lithium*. Phys. Rev., **94**, 1111 (1954).

[133] A. Gonis. *Green functions for ordered and disordered systems, Studies in mathematical physics, vol. 4*. North-Holland Elsevier science publishers B.V. (1992).

[134] J. Zabloudil, R. Hammerling, L. Szunyogh and P. Weinberger. *Electron Scattering in Solid Matter*. Springer Series in Solid-State Sciences, **147** (2005).

[135] P. Ignatiev. *Theoretical study of spin-polarized surface states on metal surfaces*. Ph.D. thesis, Max-Planck-Institute of Microstructure Physics, Halle (Saale) (2009).

[136] I. Mertig, E. Mrosan and P. Ziesche. *Multiple scattering theory of point defects in metals : electronic properties, volume 11 of Teubner-Texte zur Physik*. Teubner (1987).

[137] E. N. Economou. *Green's Functions in Quantum Physics*. Springer, Berlin (2006).

[138] R. G. Newton. *Scattering Theory of Waves and Particles: Second Edition*. Dover Publications Inc. (2002).

[139] P. Mavropoulos and N. Papanikolaou. *The Korringa-Kohn-Rostoker (KK) Green Function Method I: Electronic Structure of Periodic Systems*. Computational Nanoscience, John von Neumann Institute for Computing, Jülich, **31**, 131 (2006).

[140] A. Gonis. *Theoretical Materials Science: Tracing the Electronic Origins of Materials Behavior*. Materials Research Society (2000).

[141] R. Zeller, P. H. Dederichs, B. Újfalussy, L. Szunyogh and P. Weinberger. *Theory and convergence properties of the screened Korringa-Kohn-Rostoker method*. Phys. Rev. B, **52**, 8807 (1995).

[142] N. Papanikolaou, R. Zeller and P. H. Dederichs. *Conceptual improvements of the KKR method*. J. Phys.: Cond. Matt., **14**, 2799 (2002).

[143] R. Zeller and P. H. Dederichs. *Electronic Structure of Impurities in Cu, Calculated Self-Consistently by Korringa-Kohn-Rostoker Green's-Function Method*. Phys. Rev. Lett., **42**, 1713 (1979).

[144] R. Podloucky, R. Zeller and P. H. Dederichs. *Electronic structure of magnetic impurities calculated from first principles*. Phys. Rev. B, **22**, 5777 (1980).

[145] K. Wildberger, V. S. Stepanyuk, P. Lang, R. Zeller and P. H. Dederichs. *Magnetic Nanostructures: 4d Clusters on Ag(001)*. Phys. Rev. Lett., **75**, 509 (1995).

[146] V. S. Stepanyuk, W. Hergert, P. Rennert, K. Wildberger, R. Zeller and P. H. Dederichs. *Magnetic dimers of transition-metal atoms on the Ag(001) surface.* Phys. Rev. B, **54**, 14121 (1996).

[147] V. S. Stepanyuk, W. Hergert, K. Wildberger, R. Zeller and P. H. Dederichs. *Magnetism of 3d, 4d, and 5d transition-metal impurities on Pd(001) and Pt(001) surfaces.* Phys. Rev. B, **53**, 2121 (1996).

[148] P. J. Braspenning, R. Zeller, P. H. Dederichs and A. Lodder. *Electronic structure of non-magnetic impurities in Cu.* J. Phys. F: Met. Phys., **12**, 105 (1982).

[149] E. Y. Tsymbal. *Lectures: Introduction to Solid State Physics* (2005).

[150] H. J. Monkhorst and J. D. Pack. *Special points for Brillouin-zone integrations.* Phys. Rev. B, **13**, 5188 (1976).

[151] L. Bürgi, L. Petersen, H. Brune and K. Kern. *Noble metal surface states: deviations from parabolic dispersion.* Surf. Sci., **447**, L157 (2000).

[152] F. Reinert, G. Nicolay, S. Schmidt, D. Ehm and S. Hüfner. *Direct measurements of the L-gap surface states on the (111) face of noble metals by photoelectron spectroscopy.* Phys. Rev. B, **63**, 115415 (2001).

[153] J. A. Stroscio, F. Tavazza, J. N. Crain, R. J. Celotta and A. M. Chaka. *Electronically Induced Atom Motion in Engineered CoCun Nanostructures.* Science, **313**, 948 (2006).

[154] J. Lagoute, C. Nacci and S. Fölsch. *Doping of Monatomic Cu Chains with Single Co Atoms.* Phys. Rev. Lett., **98**, 146804 (2007).

[155] M. Zheng, J. Shen, J. Barthel, P. Ohresser, C. V. Mohan and J. Kirschner. *Growth, structure and magnetic properties of Co ultrathin films on Cu(111) by pulsed laser deposition.* J. Phys.: Cond. Matt., **12**, 783 (2000).

[156] S. S. P. Parkin, N. More and K. P. Roche. *Oscillations in exchange coupling and magnetoresistance in metallic superlattice structures: Co/Ru, Co/Cr, and Fe/Cr.* Phys. Rev. Lett., **64**, 2304 (1990).

[157] T. Uchihashi, J. Zhang, J. Kröger and R. Berndt. *Quantum modulation of the Kondo resonance of Co adatoms on Cu/Co/Cu(100): Low-temperature scanning tunneling spectroscopy study.* Phys. Rev. B, **78**, 033402 (2008).

[158] V. S. Stepanyuk, L. Niebergall, R. C. Longo, W. Hergert and P. Bruno. *Magnetic nanostructures stabilized by surface-state electrons.* Phys. Rev. B, **70**, 075414 (2004).

[159] L. Niebergall, V. S. Stepanyuk, J. Berakdar and P. Bruno. *Controlling the Spin Polarization of Nanostructures on Magnetic Substrates.* Phys. Rev. Lett., **96**, 127204 (2006).

[160] E. Cox, M. Li, P.-W. Chung, C. Ghosh, T. S. Rahman, C. J. Jenks, J. W. Evans and P. A. Thiel. *Temperature dependence of island growth shapes during submonolayer deposition of Ag on Ag(111).* Phys. Rev. B, **71**, 115414 (2005).

[161] M. Giesen and G. S. Icking-Konert. *Equilibrium fluctuations and decay of step bumps on vicinal Cu (111) surfaces.* Surf. Sci., **412-413**, 645 (1998).

[162] J. T. Li, W. D. Schneider, R. Berndt and S. Crampin. *Electron confinement to nanoscale Ag islands on Ag(111): A quantitative study.* Phys. Rev. Lett., **80**, 3332 (1998).

[163] J. Li, W. D. Schneider, S. Crampin and R. Berndt. *Tunnelling spectroscopy of surface state scattering and confinement.* Surf. Sci., **422**, 95 (1999).

[164] L. Niebergall, G. Rodary, H. F. Ding, D. Sander, V. S. Stepanyuk, P. Bruno and J. Kirschner. *Electron confinement in hexagonal vacancy islands: Theory and experiment.* Phys. Rev. B, **74**, 195436 (2006).

[165] O. Pietzsch, S. Okatov, A. Kubetzka, M. Bode, S. Heinze, A. Lichtenstein and R. Wiesendanger. *Spin-Resolved Electronic Structure of Nanoscale Cobalt Islands on Cu(111).* Phys. Rev. Lett., **96**, 237203 (2006).

[166] G. Rodary, D. Sander, H. Liu, H. Zhao, L. Niebergall, V. S. Stepanyuk, P. Bruno and J. Kirschner. *Quantization of the electron wave vector in nanostructures: Counting k-states.* Phys. Rev. B, **75**, 233412 (2007).

[167] L. Limot, E. Pehlke, J. Kröger and R. Berndt. *Surface-State Localization at Adatoms.* Phys. Rev. Lett., **94**, 036805 (2005).

[168] B. Simon. *The bound state of weakly coupled Schrödinger operators in one and two dimensions.* Annals of Physics, **97**, 279 (1976).

[169] F. E. Olsson, M. Persson, A. G. Borisov, J.-P. Gauyacq, J. Lagoute and S. Fölsch. *Localization of the Cu(111) Surface State by Single Cu Adatoms.* Phys. Rev. Lett., **93**, 206803 (2004).

[170] A. G. Borisov, A. K. Kazansky and J. P. Gauyacq. *Resonances induced by Cs adsorbates on Cu(100): Localization of image potential states.* Phys. Rev. B, **65**, 205414 (2002).

[171] V. S. Stepanyuk, A. N. Klavsyuk, L. Niebergall and P. Bruno. *End electronic states in Cu chains on Cu(111): Ab initio calculations.* Phys. Rev. B, **72**, 153407 (2005).

[172] B. Lazarovits, L. Szunyogh and P. Weinberger. *Spin-polarized surface states close to adatoms on Cu(111).* Phys. Rev. B, **73**, 045430 (2006).

[173] S. Lounis, P. Mavropoulos, P. H. Dederichs and S. Blügel. *Surface-state scattering by adatoms on noble metals: Ab initio calculations using the Korringa-Kohn-Rostoker Green function method.* Phys. Rev. B, **73**, 195421 (2006).

[174] H. J. Lee, W. Ho and M. Persson. *Spin Splitting of s and p States in Single Atoms and Magnetic Coupling in Dimers on a Surface.* Phys. Rev. Lett., **92**, 186802 (2004).

[175] M. E. Flatté and D. E. Reynolds. *Local spectrum of a superconductor as a probe of interactions between magnetic impurities.* Phys. Rev. B, **61**, 14810 (2000).

[176] D. Kitchen, A. Richardella, J.-M. Tang, M. E. Flatte and A. Yazdani. *Atom-by-atom substitution of Mn in GaAs and visualization of their hole-mediated interactions.* Nature, **442**, 436 (2006).

[177] A. Bergman, L. Nordström, A. B. Klautau, S. Frota-Pessôa and O. Eriksson. *Magnetic structures of small Fe, Mn, and Cr clusters supported on Cu(111): Noncollinear first-principles calculations.* Phys. Rev. B, **75**, 224425 (2007).

[178] R. Robles and L. Nordström. *Noncollinear magnetism of Cr clusters on Fe surfaces.* Phys. Rev. B, **74**, 094403 (2006).

[179] G. Binnig, C. F. Quate and C. Gerber. *Atomic Force Microscope.* Phys. Rev. Lett., **56**, 930 (1986).

[180] K. Teichmann, M. Wenderoth, S. Loth, R. G. Ulbrich, J. K. Garleff, A. P. Wijnheijmer and P. M. Koenraad. *Controlled Charge Switching on a Single Donor with a Scanning Tunneling Microscope.* Phys. Rev. Lett., **101**, 076103 (2008).

[181] S. Crampin. *Surface states as probes of buried impurities.* J. Phys.: Cond. Mat., **6**, L613 (1994).

[182] S. Heinze, R. Abt, S. Blügel, G. Gilarowski and H. Niehus. *Scanning Tunneling Microscopy Images of Transition-Metal Structures Buried Below Noble-Metal Surfaces.* Phys. Rev. Lett., **83**, 4808 (1999).

[183] M. C. M. M. van der Wielen, A. J. A. van Roij and H. van Kempen. *Direct Observation of Friedel Oscillations around Incorporated SiGa Dopants in GaAs by Low-Temperature Scanning Tunneling Microscopy.* Phys. Rev. Lett., **76**, 1075 (1996).

[184] K. Kobayashi. *Scattering theory of subsurface impurities observed in scanning tunneling microscopy.* Phys. Rev. B, **54**, 17029 (1996).

[185] S. Heinze, G. Bihlmayer and S. Bluger. *First-principles interpretation of scanning tunneling microscopy applied to transition-metal surfaces: Buried CuIr/Cu(001) surface alloys.* Phys. Stat. Sol. A, **187**, 215 (2001).

[186] N. Quaas, M. Wenderoth, A. Weismann, R. G. Ulbrich and K. Schönhammer. *Kondo resonance of single Co atoms embedded in Cu(111).* Phys. Rev. B, **69**, 201103 (2004).

[187] Y. S. Avotina, Y. A. Kolesnichenko, A. N. Omelyanchouk, A. F. Otte and J. M. van Ruitenbeek. *Method to determine defect positions below a metal surface by STM.* Phys. Rev. B, **71**, 115430 (2005).

[188] C. Didiot, A. Vedeneev, Y. Fagot-Revurat, B. Kierren and D. Malterre. *Imaging a buried interface by scanning tunneling spectroscopy of surface states in a metallic system.* Phys. Rev. B, **72**, 233408 (2005).

[189] I. B. Altfeder, D. M. Chen and K. A. Matveev. *Imaging Buried Interfacial Lattices with Quantized Electrons.* Phys. Rev. Lett., **80**, 4895 (1998).

[190] A. Weismann, M. Wenderoth, S. Lounis, P. Zahn, N. Quaas, R. G. Ulbrich, P. H. Dederichs and S. Blugel. *Seeing the Fermi Surface in Real Space by Nanoscale Electron Focusing.* Science, **323**, 1190 (2009).

[191] Y. S. Avotina, Y. A. Kolesnichenko, A. F. Otte and J. M. van Ruitenbeek. *Signature of Fermi-surface anisotropy in point contact conductance in the presence of defects.* Phys. Rev. B, **74**, 085411 (2006).

[192] Y. S. Avotina, Y. A. Kolesnichenko, A. F. Otte and J. M. van Ruitenbeek. *Magneto-quantum oscillations of the conductance of a tunnel point contact in the presence of a single defect.* Phys. Rev. B, **75**, 125411 (2007).

[193] Y. S. Avotina, Y. A. Kolesnichenko and J. M. van Ruitenbeek. *Magneto-orientation and quantum size effects in spin-polarized STM conductance in the presence of a sub-surface magnetic cluster.* Phys. Rev. B, **80**, 115333 (2009).

[194] M. A. Torija, A. P. Li, X. C. Guan, E. W. Plummer and J. Shen. *"Live" Surface Ferromagnetism in Fe Nanodots/Cu Multilayers on Cu(111).* Phys. Rev. Lett., **95**, 257203 (2005).

[195] J. Xu, M. A. Howson, B. J. Hickey, D. Greig, E. Kolb, P. Veillet and N. Wiser. *Superparamagnetism and different growth mechanisms of Co/Au(111) and Co/Cu(111) multilayers grown by molecular-beam epitaxy.* Phys. Rev. B, **55**, 416 (1997).

i want morebooks!

Buy your books fast and straightforward online - at one of world's fastest growing online book stores! Environmentally sound due to Print-on-Demand technologies.

Buy your books online at

www.get-morebooks.com

Kaufen Sie Ihre Bücher schnell und unkompliziert online – auf einer der am schnellsten wachsenden Buchhandelsplattformen weltweit! Dank Print-On-Demand umwelt- und ressourcenschonend produziert.

Bücher schneller online kaufen

www.morebooks.de

 VDM Verlagsservicegesellschaft mbH
Heinrich-Böcking-Str. 6-8
D - 66121 Saarbrücken

Telefon: +49 681 3720 174
Telefax: +49 681 3720 1749

info@vdm-vsg.de
www.vdm-vsg.de

Printed by Books on Demand GmbH, Norderstedt / Germany